Novel Engineering, K–8

An Integrated Approach to Engineering and Literacy

Novel Engineering, K–8

An Integrated Approach
to Engineering and Literacy

Elissa Milto, Merredith Portsmore, Jessica Watkins,
Mary McCormick, and Morgan Hynes

nsta Press
National Science Teaching Association
Arlington, Virginia

Claire Reinburg, Director
Rachel Ledbetter, Managing Editor
Jennifer Merrill, Associate Editor
Andrea Silen, Associate Editor
Donna Yudkin, Book Acquisitions Manager

ART AND DESIGN
Will Thomas Jr., Director, cover
Capital Communications LLC, interior design

PRINTING AND PRODUCTION
Catherine Lorrain, Director

NATIONAL SCIENCE TEACHING ASSOCIATION
1840 Wilson Blvd., Arlington, VA 22201
www.nsta.org/store
For customer service inquiries, please call 800-277-5300.

NSTA is committed to publishing material that promotes the best in inquiry-based science education. However, conditions of actual use may vary, and the safety procedures and practices described in this book are intended to serve only as a guide. Additional precautionary measures may be required. NSTA and the authors do not warrant or represent that the procedures and practices in this book meet any safety code or standard of federal, state, or local regulations. NSTA and the authors disclaim any liability for personal injury or damage to property arising out of or relating to the use of this book, including any of the recommendations, instructions, or materials contained therein.

Cataloging-in-Publication data for this book and the e-book are available from the Library of Congress.
ISBN: 978-1-68140-642-8
e-ISBN: 978-1-68140-643-5

Contents

Contents

APPENDIXES

Preface

Novel Engineering is an innovative approach to including engineering in K–8 classrooms. As we've worked on Novel Engineering, we have been encouraged by the excitement that students and teachers have shown for their work during Novel Engineering units. One of the things we are excited about is that students have taken ownership of their learning and are able to navigate the steps of an engineering design process (EDP)—creating functional solutions to problems they have identified in texts. Of course, students would not be able to do this without the support and freedom given to them by their teachers. Many of the teachers with whom we've spoken have reflected that their own teaching has changed because they're better able to recognize what their students are thinking and understand their work.

In the United States, K–12 engineering education has gone from relative obscurity 20 years ago to part of the national science standards today. As it has evolved, we have seen that there are different ways of incorporating engineering into classrooms. Often, K–12 engineering activities are used as a vehicle to motivate or deliver science and math content, or they are stand-alone experiences for students to learn engineering with limited connections to science and math. In addition, they often do not emulate the experiences of real-world engineering.

Professional engineers have rich contexts in which they design. They have multiple stakeholders with different needs they must translate into design requirements; they have constraints on materials, time, or solution types they need to account for and balance; and they must address regulations and ethical issues. Professional engineers in these contexts are skilled at finding problems, identifying requirements, and balancing trade-offs. Often, when we transpose engineering into K–12 settings, some of the richness and wonderful messiness of real-world engineering is lost in activities that specify the problem and all the requirements for students. Engineering curricula for K–8 students are often constrained both in problem definition and in material and solution diversity. These activities are great ways for students to engage in elements of engineering design and are often necessary with the realities of school, but they don't allow students to participate in the full design process.

Novel Engineering represents a significant shift in thinking about how engineering knowledge can codevelop with other disciplines, usually literacy but frequently social studies or another core subject. Novel Engineering works to replace real-world clients and contexts with those from literary texts to offer students opportunities to enter into engineering design practices that are messy, ill-defined, and without predetermined "correct" answers or paths through the EDP. Novel Engineering has shown a new model for engineering integration in which two disciplines are truly dependent on each other—and an approach where students are able to navigate the two successfully.

This book aims to support educators as they do Novel Engineering in their own classrooms and teacher educators as they work with preservice teachers. There are many teachers who have done Novel Engineering or have attended in-person professional development workshops, but the number of interested teachers and the range of locations make it impossible to meet with all teachers who would like to implement Novel Engineering with their students. It is our hope that through this book, we will be able to share Novel Engineering with a greater number of teachers.

This book will not only describe the Novel Engineering approach but also present case studies that allow readers to practice noticing student thinking and begin anticipating what students may do and say. In addition, this book will walk readers through the planning of a Novel Engineering unit so they can use the books they already include in their curriculum as part of a unit. By including Novel Engineering in the classroom, teachers give students the chance to engage in all steps of an EDP in a way that is personally meaningful.

Novel Engineering began in 2010 as a research project supported by the National Science Foundation. The first phase of the project was to understand what engineering looks like at the elementary school level. We examined what the engineering students were doing and what that looked like within an integrated context. As the research progressed, we also looked at Novel Engineering in middle schools and with students with language-based learning disabilities. Throughout the project, we collaborated with teachers to help us understand a teacher's point of view.

The Novel Engineering team had a broad understanding of the interaction among engineering and education research, educational tool development, and classroom implementation. The research team included individuals with backgrounds in engineering, engineering education, psychology, literacy, science, education, and special education. Teachers participating in the research portion of the project functioned as partners, offering unique insights into the team's research and implementation. This diverse team allowed us to understand what was happening with students from a multitude of perspectives.

In addition to this background experience, the team worked directly with students while they were engineering. This allowed us to see what worked and what did not work and to better understand what students were capable of doing. This understanding of classroom dynamics and our partnerships with classroom teachers helped us develop not only the Novel Engineering approach but also the professional development experience.

Overview

Novel Engineering is an integrated approach to teaching engineering and literacy in elementary and middle school classrooms. Through this approach, students use literature as the basis for engineering design challenges, drawing information from the text to identify engineering problems, considering characters as clients, and using details from the story to impose constraints as they build functional solutions to the characters' problems.

For example, students who read the book *Danny the Champion of the World* by Roald Dahl identified Danny's father falling into a pit and getting stuck as a problem and then built and tested functional models to solve that problem (i.e., to get his father out of the pit) in a way that used resources appropriate to the story's setting. As students work on the text-based engineering projects, they also engage in productive and self-directed literacy practices. Novel Engineering tasks are truly interdisciplinary efforts in which students engage in both engineering and literacy activity. One teacher said, "That kids are using problem-solving skills based on basic engineering strategies, and it's an interdisciplinary unit combining science, math, and English language arts. … I think it's great—[it] gets kids to think in a different way." Since Novel Engineering began, we've been consistently excited by the work students have done and where our partnerships with teachers have led the project.

Novel Engineering continues to advance how students interpret classroom activities and how that interpretation influences the abilities and practices students leverage for learning. In Novel Engineering, we see students navigate classroom constraints, constraints presented by the text, and constraints of the real world. Engineering education literature often talks about students' limitations or inabilities. Novel Engineering's research findings add to the conversation about young students' sophisticated abilities and knowledge, and they push for more work on unpacking the significance of context for students (Watkins, Spencer, and Hammer 2014).

Through our research, we've found that Novel Engineering benefits students in multiple ways. It provides a context for students to more deeply engage with assigned reading texts, whether they be in English language arts or history. The text serves as the basis for discourse, argumentation, and the sharing of ideas and thinking. In order to design for characters/clients, students need to make inferences and predictions based on what they've read that will influence their designs. There are many opportunities during a Novel Engineering unit for students to write, discuss what they have read, and argue for their point of view using evidence from the text.

In addition, the research conducted during the project has helped advance conceptions of what students are capable of with respect to engineering and how their capabilities are recognized in the classroom. Finally, Novel Engineering research has also advanced models and resources for preparing teachers to teach engineering, such as using videos of students engaged in engineering to develop teachers' abilities to notice when their students are showing emergent engineering skills.

Teachers have pointed out that Novel Engineering can reach students with different learning profiles—not just the "A students" who typically do well. For example, we've seen students with special needs excel at Novel Engineering; we've seen several instances of students with reading disabilities easily access a text supporting evidence for their design ideas. We've also seen students with significant organizational deficits manage the complex process of planning and realizing their ideas, which is often difficult to do when balancing multiple constraints and steps.

Wrap-Up

We know that readers will come to this book with a variety of backgrounds and experiences. It is for this reason that we imagine not everyone will read the book cover to cover and that readers may move between the chapters they feel best complement their experience levels with engineering and/or literacy. This book is broken into three sections: an overview of Novel Engineering's links to engineering and literacy, case studies, and logistical information for implementation.

We hope you enjoy reading this book and doing Novel Engineering with your students. We are continuously surprised by the amazing things we've seen students engineer and the discussions they've had about the books they read. We are sure you will be amazed by your students, as well.

Safety Considerations

Student safety is a primary consideration in all subjects, but it is an area of particular concern in science/engineering since students interact with tools and materials with which they are unfamiliar, posing additional safety risks. Teachers need to be sure that their rooms and other spaces are appropriate for the activities being conducted. That means that engineering controls such as proper ventilation, fire extinguishers, and eye wash stations are available and utilized properly. In addition, students should use sanitized indirectly vented chemical-splash goggles, safety glasses with side shields, nonlatex aprons, and vinyl gloves during all components of an investigation (i.e., setup, hands-on investigation, clean-up) in which they handle potentially harmful supplies, equipment, or chemicals. At a minimum, the eye protection provided to students must meet the ANSI/ISEA Z87.1 D3 standard.

Remember also to review and comply with all safety policies and procedures that have been established by your place of employment. Teachers must practice proper disposal of materials and proper maintenance of all equipment. The National Science Teaching Association (NSTA) maintains an excellent website (*www.nsta.org/safety*) that provides guidance for teachers at all levels. The site also has safety acknowledgment forms for each grade level. These forms are for students to review with their teachers and must be signed by parents/guardians.

Safety Notes are included in certain chapters to highlight specific safety concerns that might be associated with particular lessons. The safety precautions associated with each investigation are based in part on the use of the recommended materials and instructions, legal safety standards, and better professional safety practices. Selection of alternative materials or procedures for these investigations may jeopardize the level of safety and is therefore at the user's own risk.

Reference

Watkins, J., K. Spencer, and D. Hammer. 2014. Examining young students' problem scoping in engineering design. *Journal of Precollege Engineering Education Research* 4 (1): 43–53.

Acknowledgments

Novel Engineering started as an idea and grew into an approach to learning engineering that has been more powerful than anyone could have imagined when we first started. We are very grateful for everyone who has helped advance the project to the state where it could become a book.

We'd like to thank the following people:

- Bill Wolfson for the initial inspiration for the project and Karen DeRusha for sharing her experiences.

- All the teachers who supported the research and testing of Novel Engineering in their classrooms. Special thanks to Maggie Jackson and Anne Valluzzi for their dedication in helping to write Chapters 4 and 5 of this book.

- The rest of the Novel Engineering Leadership Team: Ethan Danahy, David Hammer, Chris Rogers, Kathleen Spencer, and Kristen Wendell.

- Our Novel Engineering Team: Chelsea Andrews, Susan Bitetti, Andy Braren, Sarah Coppola, Elise Deitrick, Tafari Duncan, Laura Fradin, Philip Gay, Daniel Haack, Caitlin Hall-Swan, Jeff Govoni, Aaron Johnson, Chip Jones, Emma Jones, Quinn Jones, Gard Ligonde, Dan Lu, Leonardo Madariaga, Kerrianne Marino, Bridget McCafferty, Lajja Mehta, Lisa Meyers, Matthew Mueller, Brian O'Connell, Victoria Portsmore, Alex Pugnali, Erin Riecker, Sarah Rosenberg, Jennifer Scinto, Jessica Scolnic, Ben Shapiro, Victoria Sims, Kathleen Spencer, Jessica Swenson, Nathan Tarrh, Jennifer Thomas, April Truong, Anne Vakkuzzi, Dan Wise, Anne Worrall, Megan Wyllie, Rafi Yagudin, and Lija Yang.

- The Novel Engineering Advisory Board: Jake Foster, Amy Tonkonogy, and Maryanne Wolf.

Acknowledgments

- Our collaborators at Washington University, St. Louis (Vicki May, Paula Smith, Melanie Turnage, Kim Weaver); at North Carolina State University (Amber Kendall); at White Mountain Science Institute (Bill Church); Anne Kaufman-Frederick, Learning Innovation and Consulting Services for Education; and the amazing engineering education fairy godmother, Liz Parry.

- The McDonnell Family Foundation for supporting Novel Engineering in St. Louis and the development of additional resources.
- Tufts CEEO Operations Team—Magee Shalhoub, Lynne Ramsey, and Alison Blanchard—for their logistical support of finances, ordering materials, and more.

This book is based on work supported by the National Science Foundation under Grant No. 1020243. Any opinions, findings, conclusions, or recommendations expressed in this material are those of the authors and do not necessarily reflect the views of the National Science Foundation.

About the Authors

Elissa Milto

*Director of Outreach, Tufts Center for
Engineering Education and Outreach*
Tufts University, Medford, Massachusetts

Elissa's background in teaching and special education led her to purse her second master's in engineering education and to become a core member of Tufts Center for Engineering Education and Outreach. As Director of Outreach, Elissa leads the center's work to provide schools, teachers, and other organizations with engineering design opportunities grounded in research. She leads the Student Teacher Outreach Mentorship Program (STOMP), a community-based outreach program; Design & Engineering Workshops; summer and afterschool programs for students and K–12 teachers; and she consults with local schools and international groups. Elissa is particularly interested in using open-ended, client-centered problems to bring engineering to elementary and middle school students and exploring ways that students with different learning styles and interests can become excited by and access engineering. Elissa has been leading the Novel Engineering project (*www.novelengineering.org*) since it began in 2010.

Merredith Portsmore

Director and Research Assistant Professor,
Tufts Center for Engineering Education
and Outreach
Tufts University, Medford, Massachusetts

Merredith Portsmore is the director of Tufts Center for Engineering Education and Outreach and a research assistant professor. She has the unique honor of being a "quadruple jumbo," having received her four degrees from Tufts University (BA in English, BS in mechanical engineering, MA in education, and PhD in engineering education). Merredith's research interests focus on how children engage in constructing solutions to engineering design problems, how teachers learn engineering in online environments, and how outreach programs affect K–12 students. Her outreach work focuses on creating resources for K–12 educators to support engineering education in the classrooms. She is the founder of STOMP, LEGOEngineering.com, and the online Teacher Engineering Education Program (*https://teep.tufts.edu*).

Jessica Watkins

Assistant Professor, Department of
Teaching and Learning
Vanderbilt University, Nashville, Tennessee

Jessica is an assistant professor of science education at Vanderbilt University. She received her PhD from Harvard University, studying the effects of undergraduate reformed physics courses. In her current research, she studies how K–16 students engage in science and engineering as disciplinary pursuits—that is, how they seek deeper understandings of the natural world and solutions to problems within it. She examines both moment-to-moment interactions and long-term dynamics of learners' engagement to understand how their disciplinary pursuits get started and progress over time. Grounding her research is the understanding that learners bring diverse, productive resources for science and that engineering and the role of educators is to help learners build on and refine these resources. Therefore, a second strand of her research is to design and study contexts in which teachers learn to attend to students' thinking in productive and expansive ways.

Mary McCormick

*Consultant, Tufts Center for Engineering
Education and Outreach*

Tufts University, Medford, Massachusetts

Mary started her career in civil engineering working as a geotechnical engineer. When she returned to graduate school, her passion was ignited for engineering education. Mary received her PhD in STEM Education at Tufts, where she was a core member of the Novel Engineering research team. Her research focused on the dynamics of students' framing (sense of the task) during Novel Engineering activities. Mary currently works as a consultant at the Tufts Center for Engineering Education and Outreach, working on research, teaching, and writing projects.

Morgan Hynes

*Associate Professor, Department of
Engineering Education*

Purdue University, Lafayette, Indiana

Morgan is an assistant professor in the School of Engineering Education at Purdue University and director of the For All: A Chance to Engineer (FACE) Lab research group at Purdue. In his research, Morgan explores the use of engineering to integrate academic subjects in K–12 classrooms. Specific research interests include design metacognition among learners of all ages; the knowledge base for teaching K–12 STEM through engineering; the relationships among the attitudes, beliefs, motivation, cognitive skills, and engineering skills of K–16 engineering learners; and the teaching of engineering.

Section I

What Is Novel Engineering?

Introduction to Novel Engineering

Chapter 1

A lot of talk in education focuses on integration—combining subjects in meaningful ways to help students learn and see how knowledge and practices cross disciplinary boundaries. Novel Engineering, which follows the trajectory in Figure 1.1, can be taught as part of an English language arts (ELA) curriculum. It has also been implemented in other disciplines. Most of our research took place in ELA classes, so that is where most of the examples in this book take place. At first glance, engineering and ELA may seem like an unlikely pair for integration.

Figure 1.1: Novel Engineering design trajectory

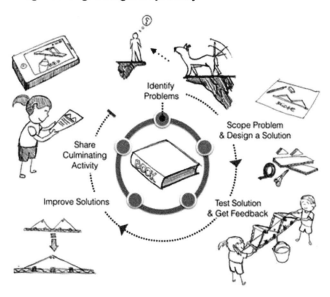

On the one hand, there's engineering, which focuses on solving problems through iterative design. Engineering also pursues solutions to problems through careful identification of needs, requirements, and iterative prototyping, testing, and revision. Literacy, on the other hand, teaches students how to comprehend and interpret text to build understanding and how to engage in discussion, both oral and written, about text.

Novel Engineering gives students the opportunity to enter into engineering design through literature, offering authentic engineering projects that do not have predetermined, "correct" answers. While working on a Novel Engineering unit, students engage in engineering by drawing on their past experiences and understandings of the world and interact with classmates about what's happening in the book and what they have built. As students work on text-based engineering projects, they also engage in productive and self-directed literacy practices, including noting key details in text, making inferences, and writing lists and other notes that support the design process. Novel Engineering projects are therefore interdisciplinary efforts in which students gain experience in both disciplines.

One of the benefits of Novel Engineering is that it allows teachers to use some of their literacy blocks for projects since part of the students' time is used to interact with the text. Novel Engineering is similar to project-based learning in that curricular goals address more than one discipline, but it is different from project-based learning in that it has a specific focus on two disciplines.

This open-ended structure leads to solution diversity among groups within the same classroom. Although there is a basic framework for doing activities, there are not specific lesson plans or scripts. We've found, through our research and interactions with teachers, that by providing a framework for activities, teachers are able to develop their own content based on books that are already part of their curriculum. In fact, teachers with whom we have worked have told us how much they appreciate that Novel Engineering values their expertise and decision-making capabilities by not giving them highly structured lesson plans. It is for this reason that we have not included lesson plans in this book. However, we do include a sample lesson guide on the book's Extras page at *www.nsta.org/novelengineering*.

An Overview of Novel Engineering in the Classroom

The best way to begin this book is to sketch out what Novel Engineering can look like in a classroom. We've seen the book *Wonder* by R. J. Palacio used in several fifth-grade classrooms and are going to present a composite of these classrooms. Although there is variety among the classrooms and students, there are many similarities. *Wonder* is the story of Auggie, a fifth-grade boy who was born with a severe facial difference and is entering school for the first time. The book begins from his perspective and then switches to include the perspectives of the other characters. The teachers have several learning goals for students that include having students think intensely about the characters and the overarching themes of acceptance and friendship. This requires students to think about multiple characters' perspectives and make inferences about their thoughts and feelings. As the teachers read the book, they pause to give students time to discuss the problems that arose and to discuss, as engineers, how they might solve those problems.

As groups are engaged in discussion, the teacher walks around the room and listens to the discussions. One group wants to address the discomfort that the main character, Auggie, feels while eating in the school cafeteria. Due to his facial structure, Auggie is very messy when he eats and feels embarrassed. As two students, Samuel and Mateo, begin to consider solutions to this problem, it becomes evident that they are drawing on details of the story and making spontaneous inferences, all in service of understanding the design context. For example, they describe how they think Auggie feels, cite specific passages in the text, and infer the reason for those feelings—all of which help them empathize with Auggie about how it might feel to be bullied. They also generate a map of the cafeteria based on setting descriptions, consider the social landscape of an elementary school, and come up with a list of foods that may be easier for him to eat in public.

The following is an excerpt of a conversation between the two students. The conversations throughout this book are numbered so that if teachers are discussing them in groups, they can use the numbers to refer to students' statements.

1. **Samuel:**   He doesn't like to eat with everyone.

2. **Mateo:**   He could just not eat in the cafeteria, maybe in a classroom with a teacher?

3. **Samuel:**   No, he is in school to be with the other kids. We need to make something so he can eat in the cafeteria. What can we ...

4. **Mateo:**   He'll be afraid people will look at him.

5. **Samuel:** We can make something that will let him eat and make it less messy.

6. **Mateo:** Okay. How can it be less messy so the food doesn't fall out? Maybe something that catches food but blocks his mouth?

7. **Samuel:** It can be like a fork but hides his mouth.

The following day, the group begins building a device that will help Auggie eat with less mess. As in most Novel Engineering classrooms, the students are provided with a list of available teacher-supplied materials when they begin to plan, which typically include a variety of cheap and recyclable materials such as tape, paper clips, cardboard, string, and cloth. A suggested list of materials is included in Appendix A (p. 223).

Samuel and Mateo propose to test their device using a range of foods, such as a yogurt, apples, and cheese. As they test their device, they are reminded by the teacher to record their findings in an engineering journal so they can share findings with the class and make changes, if needed, the following day. While sharing their findings with their classmates, the students describe their design choices and rationale, the way they tested their design, and how they intend to improve it. Samuel and Mateo want it to look as much like a traditional fork as possible so Auggie will not feel self-conscious. With that in mind, they include a small guard that helps keep food in his mouth.

In many Novel Engineering units, a writing assignment is included as part of a final culminating activity. In Samuel and Mateo's class, students have been instructed to write a journal entry as Auggie, describing how the engineering solution helped him overcome the problem. The pair of boys write about how Auggie felt less fear during lunchtime and is now able to talk to a friend at the lunch table. The students make projections about how their device would help Auggie gain confidence, which in turn would affect his life. In this example, Samuel and Mateo organically worked through an engineering design process (EDP) without being required to follow the process as a checklist; rather, they were allowed to move naturally through the steps. We will discuss the EDP used in Novel Engineering in the next two chapters.

After their first Novel Engineering experience, teachers often say that their students exceeded their expectations. In the previous example, Samuel and Mateo thought deeply about how Auggie might feel in different situations, such as eating in a school cafeteria or meeting new people. They also made inferences from the text and used their knowledge of the characters to project how different scenarios might play out. The teacher spoke with students as they worked, which provided a strong understanding of what their ideas were around the text, their design choices, and their construction of the final design.

In addition to meeting ELA goals, students worked collaboratively with partners or group members, communicating their ideas and supporting one another in the process. Most surprising to teachers, however, is the way their students act like young engineers. When engaged with the *Wonder* unit, students think critically about their designs, present evidence to support their design decisions, test their ideas, evaluate those ideas, and then iterate to improve their designs.

This example mirrors the experiences of hundreds of teachers with whom we have worked. Teachers consistently indicate that the integration of engineering and literacy is synergistic and powerful. Stories provide complex settings (engineering design contexts) and characters (clients) with real problems and needs, and the students' desire to help those characters by designing functional engineering solutions motivates a deeper reading and understanding of the texts. Most important, students become excited about what they are reading, writing, designing, and building! This excitement in turn helps them make strides in both engineering and literacy, as well as in their abilities to work together, think creatively and analytically, and communicate their ideas.

Novel Engineering provides a structure for students to do engineering while simultaneously working in the content areas. Books, short stories, and nonfiction texts can offer a broad context for engineering design problems that are complete with built-in constraints and criteria. In Novel Engineering, students read and identify engineering problems in the books or other texts, consider characters as clients, and then use details from the story to build functional solutions to address the characters' problems. An example of student-generated problems based on students' work with the book *Danny the Champion of the World* can be seen in Figure 1.2 (p. 8).

Books can range from picture books appropriate for kindergarteners to more complex novels for older students (see Table 1.1, p. 9). Although we will talk about the literacy and engineering portions of Novel Engineering as distinct tasks, students actually see them as part of the same task and bounce back and forth between them minute by minute. Including a hands-on piece is more time consuming, but one of the benefits of Novel Engineering is that it allows teachers to use some of their ELA blocks for these projects and provide time for students to interact with the text.

Figure 1.2: List of student-generated problems from *Danny the Champion of the World* by Roald Dahl

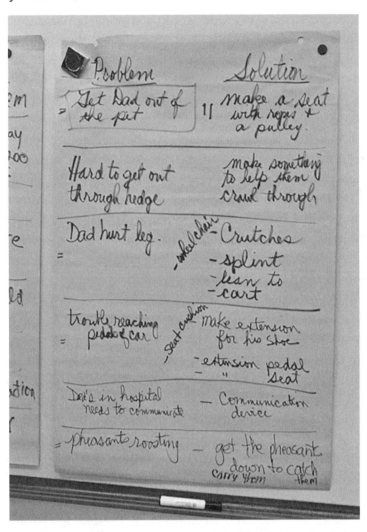

Guiding Principles of Novel Engineering

From our work at Tufts University, we've seen that students are capable of jumping into engineering projects with little guidance and that teachers can use Novel Engineering as an entry point to meet classroom goals. We've also seen that Novel Engineering provides teachers with a concrete way to attend to and respond to student thinking. These observations—along with our belief in the abilities of students and teachers—helped us formulate our guiding principles.

Table 1.1: Sample books used as part of Novel Engineering units

Title and Author	Grade Level	Lexile Level	Problems Identified by Students	Solutions Designed by Students
The Snowy Day by Ezra Jack Keats	K–2	AD500L	Keeping Peter's snowball longer	• Insulated snowball savers • Portable insulated snowball saver
Tales of a Fourth Grade Nothing by Judy Blume	3–5	470L	• Protecting Peter's pet turtle from his brother, Fudge • Preventing Fudge from getting out of his crib	• A turtle cage that prevents Fudge from getting to Peter's pet • An alarm system attached to Fudge's crib that will ring a bell when he tries to escape
A Long Walk to Water by Linda Sue Park	6–8	720L	• Nya must carry large amounts of water far distances every day • Thorns cut Nya's feet as she walks through the desert transporting the water	• A sled that can carry the water and move over rocky terrain • Shoes made of cheap materials

Novel Engineering is motivated by two guiding principles:

1. Students of all ages are capable of engineering, and their ideas can be used to inform their designs.
2. Teachers are capable of making decisions about their classrooms and their students' learning.

Rather than working from a deficit model with students, we value pre-existing student knowledge and feel that students can build on what they already know about the world as they design. Regarding teachers, the core belief that they need to be given flexibility and opportunities to make decisions about their own classrooms and students' learning means we see teachers as capable professionals who do not need "teacher-proof" curricula and should be empowered to design learning environments. Along with this flexibility is the opportunity to listen to and respond to students' ideas. When given this freedom, teachers are able to make judgments about their students' learning and decide how best to support their work.

This flexibility for both students and teachers means Novel Engineering is an open-ended approach in both how it is presented to students and how students engage in solving problems. This open-endedness means the engineering closely mirrors the real-world EDP, which is inherently messy. It also means that student engagement is elevated because students find and solve problems that are interesting to them and match their individual skills and interests. A teacher may use the same book two years in a row, but the discussions and student solutions may be very different from year to year. Although there is a trajectory that all Novel Engineering units follow (see Figure 1.1, p. 3), this serves as a path for students rather than a checklist of steps. We will talk more about the Novel Engineering trajectory in the next chapter.

Building off the first principle, students' ideas play an important role in Novel Engineering, and classroom culture should be crafted so students are comfortable sharing and acting on their ideas. We are not saying that you should let your students do whatever they want. Rather, the insight gained from understanding students' ideas gives you information about how to support students and when you can push back on their ideas. For example, there will be times when students plan to make something that is not functional or mechanically possible, such as a shrink ray, or when they want to build something that is more complicated than they have time to build.

Understanding what students are thinking will help teachers respond in a way that builds on and supports students' ideas. Rather than approaching students' work and immediately trying to improve what they are doing, teachers should take time to understand what students are thinking and why they made certain design decisions. This will help teachers respond appropriately so they can meet students where they are. Being a responsive facilitator means guiding students by asking questions and making observations so they can design realistic solutions given the available materials and time constraints.

As is evident in the dialogue between Samuel and Mateo, students are able to participate in complex discussions while navigating the EDP and thinking about the book they are reading. Teachers are often surprised and impressed with what their students can do, and students are excited to engage in hands-on, engineering design activities that do not have predetermined answers.

How Did Novel Engineering Begin?

When we started Novel Engineering, we looked at existing research from engineering, literacy, and teacher education to build on an existing concept from a local nonprofit partner, Bill Wolfson, who was using children's books to present engineering to young children. In looking at this research, we found an argument that demonstrated there was potential for design and engineering to facilitate learning in other disciplines (e.g., Kolodner 2002; Wendell and Rogers 2013) and that students' past experiences are rich resources for design projects (e.g., Portsmore 2013). This led us to think about how closely the engineering and literacy in Novel Engineering needed to be linked and how both disciplines needed to have equal value.

The Novel Engineering project began in 2010 with an interdisciplinary team of researchers and educators in engineering, literacy, education, and psychology. Our team spent the first five years of the project working with teachers and conducting research in their classrooms in rural, urban, and suburban schools. Over the course of the research, we worked with more than 500 students and filmed them as they engineered solutions to problems encountered by characters in the books they read. Research revealed what was happening as students engaged in this integrated setting and how teachers interacted with and supported their students. We documented how young students first approached engineering, how novice engineers navigated open-ended engineering design tasks, and how the integrated context influenced students' design methods (McCormick and Hammer 2016; Watkins, Spencer, and Hammer 2014). We also looked at how teachers recognize and respond to students' engineering ideas (Johnson, Wendell, and Watkins 2017; McCormick, Wendell, and O'Connell 2014; Wendell 2014).

Bidirectional Benefits

Constraints on teachers are growing and the freedom to choose what happens in the classroom is shrinking, so it may seem crazy to think of adding yet another initiative into the classroom. New initiatives must fulfill multiple functions. In our research classrooms, we saw that Novel Engineering was able to bridge several disciplines while also meeting educational standards, classroom goals, and individual goals. (In Chapters 2 and 3, we touch on how Novel Engineering aligns with *Next Generation Science Standards* and *Common Core State Standards*.) Figure 1.3 (p. 12) shows some of the benefits we've seen for students in implementing Novel Engineering in classrooms. In practice, we've found that engineering and literacy are mutually beneficial—with the text giving the engineering context and authenticity and the engineering supporting students' attention and engagement.

Figure 1.3: Bidirectional benefits of Novel Engineering

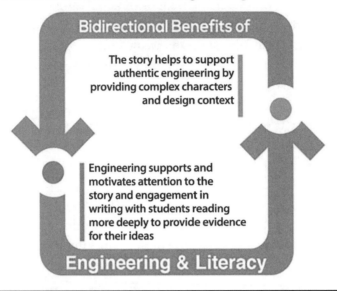

A New Kind of Resource for Teachers in Engineering and Literacy

There are many engineering curricula for young students in which students design solutions for problems rather than clients. For example, students may be tasked with building a tower of marshmallows and uncooked spaghetti. This activity may help students work on collaboration and testing skills, but it does not present students with a client or give them a context to consider as they design. With such structured tasks, there is very little solution diversity since students are working with the same materials and have been given the same constraints and criteria. Additionally, real-world engineering problems are not as neatly packaged as this. Professional engineers must sift through lots of information to figure out design criteria and constraints they need to address as they plan a design. Novel Engineering is unique in that students get to experience the messiness of engineering and have the chance to scope the problem and empathize with their clients as part of the design process.

Our research has shown that students are able to navigate the EDP without explicit directions. As we noted previously in this chapter, most elementary engineering experiences provide students with well-defined problems and a structured path through the EDP. This obviously results in all students arriving at similar solutions. In Novel Engineering, students define the engineering problems themselves as they design functional solutions based on their own ideas. As they work, they make design decisions and refine their ideas based on evidence from the book, feedback from tests, and feedback from peers. In addition to meeting standards and the goals of the classroom, Novel Engineering allows students to take ownership of a project and tackle challenging problems while working collaboratively (see Figure 1.4). This is in line with what has been outlined by the National Academy of Engineering and National Research Council (2009); it's important for all K–12 students—not just those taking engineering courses—to develop engineering habits of mind.

For the literacy aspect of Novel Engineering, we have found that students productively engage with text in a variety of ways that align with *Common Core State Standards*. They take the perspective of characters and note relevant aspects of the physical setting as they plan and evaluate their designs. Spontaneous discussions emerge as students wrestle with unfamiliar concepts and vocabulary in an effort to better inform their designs. These discussions lead to students constructing an informed interpretation of the text.

Figure 1.4: Students working on a communication system for the main character in *The City of Ember* by Jeanne DuPrau

Designed for Educators

Novel Engineering is designed to be a flexible approach for teachers. Though we recommend a semistructured flow of activities, there is no one "correct" way to do Novel Engineering, nor is there a set curriculum for teachers to follow. Novel Engineering works with most trade books and allows students to work on academic objectives identified by each teacher or school. Many teachers have found that Novel Engineering meets their academic objectives in a range of subject areas, from ELA to social studies to mathematics. Teachers have also said the Novel Engineering approach builds on their experiences and expertise rather than having them learn a completely new curriculum that does not necessarily work with the other curricula and structures that are already in place in their classrooms. Novel Engineering works well with existing curricula and plans since teachers get to choose the text and direction of student work; they can use texts students already know and are comfortable with rather than unfamiliar books from an assigned list.

Teachers have also noted that Novel Engineering helps them develop a classroom culture that includes productive class discussions and peer critique. Even though Novel Engineering requires teachers to have a basic understanding of engineering and the EDP, it does not require them to have a formal engineering background. Instead, it builds on their unique disciplinary backgrounds and classroom experiences. When working directly with teachers, we have them do a Novel Engineering unit to get a feel for the process their students will be undertaking. This book, then, is designed to help teachers gain familiarity with key aspects of engineering and help structure the process of presenting engineering to students.

Finally, Novel Engineering helps teachers look at their students in new ways. Many teachers have said that it helped them step back and notice what their students were doing and thinking. This more critical look at student thinking helps teachers understand why students make certain design decisions (related to both the text and the mechanics of the design) and why they have certain interpretations of the text. This more informed view of student thinking enables teachers to respond more appropriately to what students are doing and helps them guide their next moves. Teachers have also said that Novel Engineering helped them notice new things about their students, such as strengths or places for improvement, that did not come out in other classroom activities.

Overview of This Book

We wrote this book to share what we have learned from doing Novel Engineering in more than 100 classes. This book will walk teachers and teacher educators through the Novel Engineering approach, show concrete examples of what students may say and do, and prepare teachers to implement the approach in their classrooms. Although it's a simple concept, Novel Engineering requires teachers to anticipate what their students may do (say, think, and design), listen to their ideas, and set up structures that support students' work while meeting academic objectives.

This book is divided into three sections. Section I describes the Novel Engineering approach in more detail, what engineering looks like in young students, and how literacy and engineering support each other in project-based work. Section II consists of case studies that prepare you to lead a Novel Engineering unit and support your students as they practice being engineers. To this end, you will learn to recognize students' engineering skills and respond to student thinking. Section III includes practical elements that will be helpful in planning and implementing a Novel Engineering unit.

You do not need to read this book straight through; rather, you can jump from chapter to chapter. We definitely recommend starting with Chapters 1, 2, 3, and at least two of the case studies. After that, browse the sections you feel are most relevant to your specific classroom needs. We believe this book will help you respond to your students' work and allow them to have an authentic engineering experience as they play around in the true messiness of engineering. It will also provide students with an opportunity to be creative and follow their own ideas, taking ownership of their learning.

Visit
www.novelengineering.org
for additional Novel
Engineering resources!

References

Johnson, A. W., K. B. Wendell, and J. Watkins. 2017. Examining experienced teachers' noticing of and responses to students' engineering. *Journal of Pre-College Engineering Education Research (J-PEER)* 7 (1): 25–35.

Kolodner, J. 2002. Facilitating of learning design practices: Lessons learned from an inquiry into science. *Journal of Industrial Teacher Education* 39 (3): 9–40.

McCormick, M. E., and D. Hammer. 2016. Stable beginnings in engineering design. *Journal of Pre-College Engineering Education Research (J-PEER)* 6 (1): 45–54.

McCormick, M., K. B. Wendell, and B. P. O'Connell. 2014. Student videos as a tool for elementary teacher development in teaching engineering: What do teachers notice? (Research to practice). *Proceedings of the 2014 ASEE Annual Conference & Exposition*, Indianapolis, Indiana.

National Academy of Engineering and National Research Council. 2009. *Engineering in K–12 education: Understanding the status and improving the prospects*. Washington, DC: National Academies Press.

Portsmore, M. 2013. Exploring first grade students' drawing and artifact construction during an engineering design problem. In *"Show me what you know": Exploring representations across STEM disciplines*, ed. B. M. Brizuela and B. E. Gravel, 208–222. New York: Teachers College Press.

Watkins, J., K. Spencer, and D. Hammer. 2014. Examining young students' problem scoping in engineering design. *Journal of Pre-College Engineering Education Research (J-PEER)* 4 (1): 43–53.

Wendell, K. B. 2014. Design practices of preservice elementary teachers in an integrated engineering and literature experience. *Journal of Pre-College Engineering Education Research* 4 (2): 29–46.

Wendell, K. B., and C. Rogers. 2013. Engineering design-based science, science content performance, and science attitudes in elementary school. *Journal of Engineering Education* 102 (4): 513–540.

Websites

Common Core State Standards: *www.corestandards.org*
National Science Foundation: *www.nsf.org*
Next Generation Science Standards: *www.nextgenscience.org*
Novel Engineering: *www.novelengineering.org*

Book Resources

The City of Ember; DuPrau, J.; Age Range: 5–12; Lexile Level: GN520L
Danny the Champion of the World; Dahl, R.; Age Range: 8–12; Lexile Level: 770L
Wonder; Palacio, R. J.; Age Range: 8–12, Lexile Level: 790L

Engineering to Novel Engineering

Chapter 2

As we said in Chapter 1, Novel Engineering integrates engineering and literacy in a way that is mutually beneficial. This chapter looks at engineering in two ways. The first part takes a deep dive into what engineering is by exploring what engineers actually do. The second part examines how we can think about students doing engineering and, specifically, how they engage in engineering within Novel Engineering.

Part 1: In Pursuit of Solutions: What Is Engineering?

We often talk to students about science in the world around us, such as when we uncover new genetic information or catch a glimpse of a distant star. Though both science and engineering involve observation and analysis, science is focused on learning about the world and creating coherent explanations for the phenomena we see. By contrast, engineering tackles problems in the world and how we, as humans, create solutions for them. It leverages science, mathematics, psychology, and many other disciplines to inform those solutions. We see the output of engineering all around us—in cars, bridges, computers, gasoline, clean water, electrical power, chewing gum, medical technology, and more. But how do these products and processes come to be? How does engineering help produce these products and processes? What do engineers do to create them?

To explore engineering, let's look at a specific case of an engineer tackling a real-world problem and think about how he designed a solution and what knowledge and practices were needed to engineer that solution.

An Engineering Example: The Freedom Chair

Amos Winter, now a professor at Massachusetts Institute of Technology (MIT), was a graduate student in mechanical engineering visiting Tanzania in 2005. He was working with a nonprofit group studying how wheelchairs were used and distributed in the developing world. As a mechanical engineer, he was struck by the mismatch between a traditional wheelchair and the lack of paved roads, ramps, and transportation options for individuals in wheelchairs. He saw this was a problem that he, as an engineer, could do something about. During his initial visit, he met people who would have been able to work but were unable to get from their home to their place of work because the roads and terrain couldn't be traversed by a traditional wheelchair. He talked to them about where they lived, where they worked, and what kind of distances and terrain were involved.

As Amos pondered this problem, he began to sift through existing mechanical solutions he thought might help. He thought about gear trains and levers. He also researched some of the existing solutions. Hand-powered, tricycle wheelchairs were an option, but they weren't a viable solution for many of the people Amos met since they couldn't easily get themselves on and off of the trike wheelchairs. Mountain bike–like wheelchairs were another option, but they were far too expensive for the developing world. As Amos continued to investigate the issue, he found that the problem he had identified was widespread—more than 40 million people in the developing world needed wheelchairs but didn't have access to them. Wheelchairs are expensive and those that did exist had trouble maneuvering over the local terrain.

While thinking about the problem and possible solutions, Amos also had to consider the unique needs and challenges of different countries and settings. In Tanzania, he talked to mechanics and other people who fixed bicycles about the resources, parts, and tools that were available to them. Special parts were hard to obtain for local mechanics, but many of them repaired basic bicycles, so parts such as wheels, gears, and chains were available and affordable.

From his first trip, Amos developed a set of criteria and constraints that would guide him as he outlined a solution. In engineering, criteria generally measure success (i.e., describe what the user or designer wants to happen), and constraints are the strict requirements and/or limitations on the design imposed by the client, materials, or context. Amos's constraints were that the design had to be usable by individuals who were confined to a wheelchair and that it would help them travel the distance needed to reach a job. The developing world dictated the criteria that the design must be affordable and able to be easily repaired locally. The key distinction between criteria and constraints are that the former include things that can be measured and help designers evaluate design features/

solutions, and the latter are requirements that must be met for the design to be viable. With these criteria and constraints, Amos was able to evaluate various possible solutions (see Table 2.1).

Amos spent time thinking about ways to make the mountain bike wheelchairs that already existed more affordable. As he was brainstorming down that path, a relatively simple concept came to him as he thought about harnessing the abilities of the person sitting in the chair. He combined levers and gears to create a system in which users could change the torque by adjusting a lever connected to a gear train. This design put the complexity of changing the lever length in users' hands, which allowed them to move between more power or speed, depending on the specific terrain.

This simple idea had merit since it could be made from bicycle parts, making it affordable and locally repairable. Amos and his group at MIT (Global Research Innovation and Technology) set about making their first prototype, which they called the Freedom Chair (FC). They brought their prototype to Tanzania, Kenya, and Vietnam in 2008 and tested it with local wheelchair users. That first prototype was, in their own words, "terrible."

Amos and his group had an innovative and simple concept for moving across rough terrain, but there were additional criteria and constraints they hadn't considered enough. The chair was much heavier than a traditional wheelchair (a whopping 65 pounds versus 35 pounds for a traditional chair), making it hard for users to maneuver and hard for others to help (e.g., loading it on a truck or bus). The initial prototype was also much wider than a traditional wheelchair, making it difficult to use indoors. In the United States and other developed countries, a wheelchair user might have more than one chair for different uses. However, in the developing world, users need a single chair that can be used outside and inside and that can fit through doorways, allow them to get close to a table, and let them get into small spaces (so they can access the bathroom independently, for example).

Table 2.1: Constraints and criteria for wheelchair design

Constraints	Criteria
• Must be usable by individuals who are 100% wheelchair bound • Must help individuals travel 2–3 miles (3–5 km) per day over varied terrain	• Affordable • Locally repairable

The initial testing of their first, "terrible" prototype led to a revised set of criteria and constraints that helped them improve their solution. The FC team wasn't deterred by the lack of success they had with their first prototype; rather, they viewed it as an opportunity to learn. Engineering problems are complex, and design failure is expected. Failures provide opportunities for engineers to identify ways their design can be improved, issues they haven't yet considered, and ways in which people might use their product that they perhaps didn't anticipate.

Over the next few years, Amos and his group persisted as they iterated on their design by constructing prototypes and testing them with users. They took careful measurements of their users, who were, on average, shorter than many of the design team members (who had been the original testers). They changed the width of the seat and the length of the lever arms to better match the body dimensions of the people who would actually use the wheelchair. They also measured the users' arm strength and examined where on the chair the greatest forces and loads were. They investigated lighter materials and different frame configurations to reduce weight but maintain the necessary strength and stiffness for rigorous daily use.

As with most engineering solutions, Amos's team faced decisions where there were trade-offs among competing criteria. In order to reduce the weight of the wheelchair, the team had to consider more expensive materials or risk losing some functionality or durability. Balancing trade-offs often requires engineers to make difficult decisions to improve one criterion at the expense of another.

As new prototypes were developed, the team collected information to help guide future design decisions. The team tracked users' heart rate and oxygen use to help them choose the lever length and gear ratios that gave users the best low gear for rough terrain and high gear for smooth terrain. Learning from their mistakes, the team tracked user satisfaction with the chair in different environments, since it was essential for users to like and want to use the chair. There were multiple iterations of the chair tested in India, Guatemala, and East Africa, and each one generated a new iteration of the FC design.

As they finalized their design, there began another iterative process with the company that would manufacture the chair. Amos and his team needed to communicate their design and manufacturing ideas via computer models and parts lists. As they worked with the manufacturer, they encountered new criteria: affordability and ease of manufacturing. As a result, they had to change some aspects of the design so it would be easier to produce. Amos and his group also partnered with nonprofit organizations to raise funding to buy chairs for the users. The FC finally went into production in 2012, more than four years after the first prototype was tested, and to date, more than 2,000 individuals have received FCs in the developing world.

Looking at the FC's journey from when Amos first identified the problem to it when it was being mass produced and distributed, we see a nonlinear journey—with steps forward and backward—that used knowledge from multiple disciplines. The design team applied knowledge of science with respect to forces and mechanical advantage, economics, human anatomy and ergonomics, communication, and even human psychology. In the end, the FC was successful because of the team's attention to and care for users and their needs. In addition, the close partnerships with organizations and individuals helped ensure the design team truly understood the criteria and constraints of the problem, which they iterated on a number of times.

Engineering Design Practices and the Design Process

One way we talk about engineering is by applying an understanding of the environment in the pursuit of solutions to problems—solutions in the form of new objects, systems, or processes. The Freedom Chair is an example of engineering design that moved from an identified problem to a new object. In that example, we can see how Amos and his team progressed through multiple practices of engineering design.

When we talk about this collection of engineering practices, they are often referred to as an *engineering design process* (EDP). In professional engineering, there have been many attempts to represent this process with models that try to capture how engineering practices are invoked in the real world. They emphasize formative steps to gather information and understand the problem and then additional steps that speak to the planning, building, and testing of solutions. Figure 2.1 (p. 22) shows a simplified model of an EDP.

The process moves from identification of a problem (**Identify Problem**), which includes trying to fully understand the issue to sharing a final design with an audience (**Share**). To move between these steps, the process involves preliminary work to figure out what idea to pursue as a solution (**Research**). Research can include speaking to stakeholders and looking at past solutions. From there, engineers **Brainstorm** and generate more possible ideas. They then select an initial idea from the brainstormed options and outline design features and what needs to happen to make the proposed solution a reality (**Choose and Plan**). Engineers do not immediately begin constructing a final design but instead **Create** a prototype. They then **Test** their design, taking the successes and failures into account as they **Redesign** the prototype.

The design process isn't linear; engineers often go between steps as we saw in the FC example. Amos and his team moved between steps, sometimes repeating steps as they moved from a concept to a chair that could be manufactured on a large scale. They realized that although they had gathered some information

Chapter 2

Figure 2.1: Simplified model of the EDP

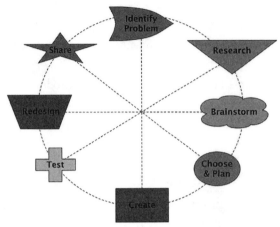

Source: Portsmore 2010.

from wheelchair users, they needed more information about specific people, places, and resources before they could design a chair that made sense for users in developing countries. However, they did not contain their fact finding to speaking with potential users; they also looked at how other people solved similar problems.

Even when the team constructed a feasible solution based on the identified constraints and criteria, they made more design refinements as they worked to manufacture the chair. Each refinement required the team to make difficult trade-off decisions whereby they made the chair more expensive to improve durability or ease of use. These sorts of trade-off decisions were guided by the team's evolving set of criteria and constraints. We can look at the engineering design practices and how they factored into the FC design process (Table 2.2).

Habits of Mind

As we look at the FC case, Amos and his group engaged in thinking practices that aren't captured by the EDP model. We frequently talk about these as the engineering habits of mind: creativity, optimism, persistence, systems thinking, conscientiousness, and collaboration. These habits of mind are an integral part of how engineers work, both independently and with a team. These are all skills that are also helpful to non-engineers, including your own students.

Creativity involves looking at the world in new ways and imagining solutions to the problems you see. Levers and gears are some of the oldest forms of engineering, but Amos and his team combined them in a new way for the FC.

Optimism is core to engineering thinking since engineers consistently look for ways to improve the world and believe good ideas from anywhere. We see this

Table 2.2: Engineering design practices in the Freedom Chair

Practice	Examples From FC
Identify Problem	Problem of transportation for disabled individuals in the developing world is identified.
Research	Research is done into how far users would travel and how chair would be prepared.
Brainstorm	There's an initial brainstorm of possible ideas to make the mountain bike wheelchair more affordable.
Choose and Plan	Amos chooses one idea to begin working on.
Create	Team creates initial models to try with users.
Test	Chair is tested with users in Guatemala, El Salvador, and India.
Redesign	Test results/user feedback are used to improve design.
Share	Computer models and descriptions for manufacturing are created.

in the FC example because, even though there are many wheelchair solutions on the market, the team knew they could add value to the world with their ideas.

Persistence is necessary for engineering because ideas fail and iteration is necessary. We have observed the FC team going through multiple trials and iterations as they unpacked new issues and worked to improve their design so it would be something that could be widely used and accepted.

Systems thinking, as an engineering habit of mind, means recognizing that the world is made up of many connected systems and problems. The FC team engaged in systems thinking as they thought about how their chair would work in a real-world environment (Would users have different chairs for different uses?), how it could be repaired (What skills and parts do local mechanics have?), and how it would be manufactured (How can it be made efficiently and affordably?).

Conscientiousness in engineering means carefully considering ethical issues and the effects of designs on people's lives. The FC team worked with a population that couldn't afford to have design failures that further damaged their health and well-being. They therefore took that into account as they thought about safety and how much testing they would need to ensure the chair was safe for individuals who already had difficulty with mobility.

Collaboration is a key engineering habit of mind since engineers need to communicate with their codesigners, listen to stakeholders, and share ideas. Very rarely do engineers work alone. Although the FC started with Amos's own ideas, he assembled a team to design it that had multiple strengths in engineering. The team also considered their clients to be codesigners and stakeholders.

Knowledge in Engineering

Thus far, we have talked about the thinking and the practices of engineering, but we haven't talked about the knowledge involved. Engineering involves an understanding of numerous other disciplines. For example, it leverages mathematics and science knowledge, which we saw throughout the FC design process. The FC team invoked their understanding of how gears and levers work and how they can be used to change the force a user applies. They also used knowledge of steel and shapes to figure out how to decrease the weight of the wheelchair and maintain the necessary strength and stiffness. Models were employed to predict how certain parts of their design would behave, and mathematics was used to perform calculations about gearing, torque, and weight. Statistics helped the team summarize data about users and their feedback to inform their design decisions.

The team also had to understand their users' preferences and the human psychology related to choosing to use or not use the chair. They applied many other types of knowledge—economics, supply chain management, manufacturing, communication, anatomy, and more—to ensure their solution would be successful along many dimensions. Although professional engineers understand multiple disciplines, they work in teams composed of members who are experts in different areas—some of which are not math or science related. For example, the FC was ultimately distributed in the developing world, so they also had to have knowledge of local culture, customs, and legal requirements.

Engineering Design as a Human Endeavor

Though we have described the richness and complexities of engineering in the FC example, engineering is not always presented this way. In fact, it is often described as a profession focused on the application of mathematics and science knowledge to achieve new solutions. There is a perception that, to become an engineer, you must love mathematics and science and excel in these subjects in school. One issue with this perspective is that it can limit the types of people who want to pursue engineering. It is true that many engineers fit into this mold, but one does not have to share these passions to become an engineer.

Engineering requires a diversity of skills, experiences, knowledge, interests, and perspectives. Unfortunately, the culture of engineering (both the profession and the college major) has been dominated by a homogenous group of people, and the lack of diversity in engineering may be limiting possible solutions and creating an inequitable environment for women and underrepresented minority groups. The 2008 book *Changing the Conversation: Messages for Improving Public*

Understanding of Engineering, published by the National Academy of Engineering, notes the importance of seeing engineering as a profession that aims to make the world a better place by helping people, animals, and the environment. The FC example is a clear demonstration of how Amos and his team aimed to make the world a better place by using their engineering skills.

A broader perspective of engineering also brings into view a more humanistic approach to the profession that aims to improve the quality of life for all people. In taking this view, you may see engineers using a much broader knowledge base that extends well beyond traditional mathematics and science knowledge and into the humanities and social sciences knowledge required to deal with ill-structured problems. One way to view the work of engineers is from the perspectives of the people they are engineering *for*, *with*, and *as*. Engineering *for* people was the central focus of the Freedom Chair, and engineers almost always do their work *with* people on project teams. Indeed, most engineered solutions require a wide variety of expertise that one person is unlikely to have on their own. Engineers also do their work *as* people, bringing their own perspectives, values, beliefs, and biases to the problem they are solving.

The FC project epitomizes this kind of engineering. Amos and his team carefully considered their clients and looked for ways to move beyond their own experiences to better understand the life of someone with disabilities in the developing world.

Part 2: Engineering for Children

The products and processes that engineers produce are an essential part of life, but why would we engage children in engineering?

In the United States, a number of factors have motivated the push to include engineering in K–12 education. The initial impetus for this movement stems from concerns about the workforce and the number of students pursuing engineering degrees (American Society for Engineering Education 1987; National Academy of Engineering 2005; National Science Foundation 2000). With flat or declining enrollment in engineering programs in colleges and universities, attention has been focused on the pipeline of students from K–12 to higher education. There was concern that students weren't engaged in the proper math and science courses to prepare them to pursue engineering degrees. There was also a realization that many students, from kindergarten through college, had little idea about what engineering was or what engineers did. These concerns and realizations were the reason for a surge in attention to engineering for children in the mid-1990s. This surge included funding for research and curriculum development centered around engineering for children. It also spawned outreach activities by universities and engineering companies that promoted engineering through awareness events and hands-on activities. There are a number of great resources that resulted from this initial concern (e.g., TeachEngineering, Engineering is Elementary).

However, even though the workforce and economy are still important factors, the motivation behind K–12 engineering education has broadened to include technological/engineering literacy for all (National Academy of Engineering and National Research Council 2009; NRC 2012; Pearson and Young 2002). This was most recently highlighted by the inclusion of engineering practices in *Next Generation Science Standards* (*NGSS*; NGSS Lead States 2013). *NGSS* assert that engineering practices are as essential for students to learn as science practice. The rationale for the inclusion of engineering practices is that the combined study of these practices will prepare students for solving major societal challenges (NRC 2012). Educators have also noticed that engineering problem-solving contexts can provide diverse opportunities to both apply and learn new ideas from a variety of school subjects (e.g., mathematics, science, literacy).

What Does Engineering for Children Look Like? What Should It Look Like?

Since the 1990s, we've seen an explosion of K–12 engineering education curricula and activities. These activities try to introduce the discipline of engineering to children. How can children do what professional engineers do? There has been much work to develop models of EDPs that use simpler, age-appropriate words, and there are lots of activities and curricula that engage students in simple engineering activities. In many cases, these activities specify a problem, requirements, and constraints. An often-used challenge is the spaghetti tower, in which students have a fixed amount of uncooked spaghetti and need to use it to build a tower that is as tall as possible. We've also seen robotics and robotic competitions grow in popularity. In order to solve a challenge, such as retrieving a box from a particular location, students need to use programming and mechanical design knowledge to build a robot that can follow a course and activate a component that will lift the box.

These challenges are all wonderful ways to engage students in portions of an EDP. As engineering for children continues to evolve and proliferate in schools, we want students to experience all the real practices, tensions, and challenges of authentic engineering design by having them not only engage in the aforementioned problems but also do the work of identifying the problem and determining the requirements and constraints. If we want students to be the next generation of innovators, they need to know how to find problems and solve ones they are given.

In addition, students are often referred to as "little scientists" and "little engineers"; indeed, research supports that even very young children have abilities for inquiry and problem solving that are similar to those of professional scientists and engineers. However, the discussion about these abilities often focuses on children's deficits—that is, the ways in which their practices are unlike those of adults. For example, researchers in science education have looked at the ways students fail to systematically test when faced with complex problems with multiple variables. In engineering, there has been a focus on how novice designers fixate on a single idea, whereas advanced designers generate multiple ideas before choosing one to prototype. In response to these deficits, recommendations have been made by researchers to scaffold and structure the learning environment to force students to engage in idea generation. An example is a worksheet that requires students to draw multiple ideas before they are given access to building materials.

There is merit in some of these scaffolds, but we have taken a different approach. As discussed in Chapter 1, we feel that students are capable of using what they know about the world to engage in engineering design. As a result, we work to create contexts in which students experience engineering tasks that are complex and ill-defined and necessitate more sophisticated engineering practices. We look at not where children lack abilities but where the learning environment could be modified to help them engage in different practices. We've seen this be productive in many contexts where students identify problems in their classroom, community, or home. A problem nested in a context that students care about allows them to bring their own knowledge to bear on the problem, care about who or what it will affect, and be able to reason through different design ideas. As educators, we look for opportunities where students can authentically engage with engineering.

Novel Engineering as Engineering

As we think about compelling engineering problems for children, we would love for them all to be as rich and complex as the FC example or representative of a real problem in the students' own communities. This allows students to experience engineering as a human endeavor. We want to offer students the chance to interact with the end users to better understand their needs and to visit the affected environment to assess the affordances and limitations of the specific setting. We want students to be powered by their own personal preferences and experiences as they select and solve problems so they can know the satisfaction and complications of designing with empathy.

However, these goals can be a challenge in classrooms limited by time, resources, and space. Novel Engineering addresses these limitations as students experience empathy, understanding, and a complex investigation of clients

and context through books. Moreover, Novel Engineering helps students see connections to the bigger picture and how engineering takes place in diverse, human contexts as opposed to a myopic focus on robots, cars, buildings, or other technologies.

Figure 2.2 shows the Novel Engineering design trajectory we've seen most classrooms follow. This is similar to the EDP we spoke about earlier in this chapter. In Table 2.3, we've included a description of each step and how it maps onto the EDP. Novel Engineering also allows teachers to address *Next Generation Science Standards* (see Appendix B, p. 224).

Novel Engineering projects elicit students' science ideas, but these units are not focused on addressing core science concepts. Instead, we focus on how to support students' development of engineering and literacy skills. It is for this reason that we do not spend much time in this book discussing how to support students' science ideas. In Chapter 7, we talk about how Novel Engineering units can be useful launching points to explore students' science ideas that emerge from their projects.

An Engineering Example: *A Long Walk to Water*

Let's look at a Novel Engineering experience with a group of middle school students in New England who read *A Long Walk to Water* by Linda Sue Park. The

Figure 2.2: Novel Engineering design trajectory

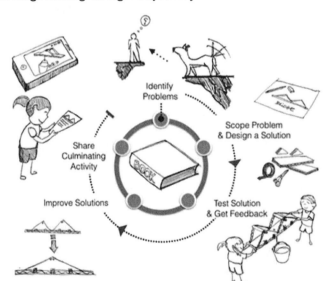

Table 2.3: Comparison of engineering design steps and Novel Engineering processes

Engineering Practice	NE Trajectory Step	Example Behavior from NE
Identify Problem	Identify Problem	Students and teachers identify problems within the book that can be solved by engineering.
Research	Scope Problem	As students read the book, they learn information about the character and setting that can inform their engineered solution.
Brainstorm	Scope Problem	Students consider various possible solutions, refining their ideas and considering constraints imposed by the book and design criteria.
Choose and Plan	Design a Solution	In their groups, students determine which of their ideas they want to pursue that makes the most sense from the book AND that they can prototype with the available materials.
Create	Design a Solution	Students construct solutions.
Test	Test Solution and Get Feedback	Students get feedback at multiple points in the design process by physically testing their designs and by presenting to the class for peer feedback.
Redesign	Improve Solution	Students take the information from their testing to make changes to their design.
Share	Share Culminating Activity	Designs are typically shared in the context of the story. How does the solution work and how would it impact the story or characters?

book tells two intertwined stories set in Sudan during that country's civil war. One story focuses on a fictionalized interpretation of the real-life experiences of Salva Dut, who was one of the Sudanese "lost boys" who ultimately started a non-profit to bring water to Sudan. The other story is fictional and focuses on a young girl, Nya, whose family is struggling to find water and safe shelter in Sudan.

The group of students examines Nya's story, which describes her role in her family's daily quest for water. The teacher decides to guide students to find a problem to address in Nya's story. Nya walks multiple hours each day through rough and dangerous terrain with her younger sister to get water. Depending on where her family is living, she must also try to get water from muddy sources.

Though the students can't meet Nya in person, the book allows them to learn about Nya and her world.

They learn that she is a hard-working member of her family, whom she cares about deeply. They hear about her struggles—including trying to manage her younger sister while walking to and from the water, how her feet hurt from walking over rocks and thorns, and how her sister becomes ill from drinking contaminated water. Students are asked to engineer for Nya. To build more understanding and empathy for her, the teacher has students carry multiple gallons of water around the room for a few minutes, which is more difficult than students anticipated.

Students then unpack ways in which they could act as engineers for Nya. They pull out problems from the text, such as her lack of shoes, the heaviness of the water, the heat, and the challenge of bringing along her little sister (see Table 2.4).

Acting as engineers, the students pick a problem to focus on. Each team identifies a problem that aligns with their own interests. One group works on ways to make a rolling wagon Nya could use to carry a larger amount of water and reduce the number of trips to the water source. Another group is excited by the idea of making shoes and chooses to construct sandals that could protect Nya's feet from the thorns and rocks along the walk.

As students work, they share their preliminary designs with their classmates as part of a presentation. As each group presents, the other students think about the book and character, consider who they are engineering for, and ask questions. Would the rolling cart be able to handle the bumpy roads described in the book? Will the shoes work in the mud by one of the water sources? Would Nya's family be able to transport the wagon easily from place to place as they move?

The students presenting then consider their classmates' questions and suggestions. The cart group looks to see if they can improve the wheels by adding thicker treads made of foam. The sandal group discusses how to make the shoes more resistant to mud by covering the shoes and laces with duct tape. These aren't generic solutions for anyone traveling to get water; students are designing specifically for Nya.

To further improve their designs, students test them in the classroom. As many are similar in age to Nya, their physical size and strength help them understand how she might interact with the design. The cart group finds that the narrow handle hurts their hands, since all the weight of the cart is focused in one spot. In response, they make the diameter of the handle wider with a piece of PVC

Table 2.4: Students' identification of problems and possible engineering solutions from *A Long Walk to Water*

Problem or Challenge in the Book	How did the characters deal with the problem? Was there a solution in the book?	Is there a better solution? Could you engineer a solution?
Thorns and no shoes	Nya didn't do anything about it and just took the pain.	She could put on shoes.
Timing	Nya had to wake up early every day and spend half of the morning walking to the water and back.	She could wake up really, really early in the morning so she could come back and have time for herself.
The heat	Nya ignores the heat and keeps walking.	She could wear extra clothes and wear something on her head for the heat.

pipe. The sandal group finds that the shoes don't stay on the heels of their feet and think of different ways they could design and build additional straps.

In the end, students share their designs with the full class. They discuss what they would improve or change if they had more time or were going to produce their designs to send to Nya. The cart group discusses the limitations of cardboard materials. They would like to make a collapsible cart that could be easier to carry when Nya's family moves to a new place. However, to keep the cardboard together, they used tape and can't enact their idea for making the cart collapsible. The sandal group would like to include more padding to make the shoes more comfortable.

While the students are building, conversations vacillate between those specific to their designs and what life is like for Nya. Students ask many questions about why Nya doesn't have access to water from a tap like they do. In the book, much of her story unfolds via Salva's story, which depicts the ongoing civil war that forces him to flee to refugee camps. Students discuss ways that professional engineers could help and how the country's political status makes it challenging for technology to be a solution. The book concludes with Salva building a nonprofit that can navigate the political landscape in Sudan and bring an engineered solution (community wells) to people, including one in Nya's village.

Looking at Engineering in This Book

The power of Novel Engineering to engage children in engineering that leverages multiple sources of knowledge is seen throughout this book, particularly in the case studies. You'll see that in some instances, students are solving grand challenges that could change the course of the whole story. However, equally powerful is students' engagement in smaller challenges where they are solving a problem that makes a character's life easier (e.g., a pair of shoes for Nya). Though Nya's story is linked to a true story, you'll see in the case studies that many types of stories, both fiction and nonfiction, help students engineer for characters.

Safety Notes

1. Wear safety goggles or glasses with side shields during the setup, hands-on, and takedown segments of the activity.
2. Use caution when using hand tools that can cut or puncture skin.
3. Use only GFI-protected circuits when using electrical equipment, and keep away from water sources to prevent shock.
4. Secure loose clothing, remove loose jewelry, wear closed-toe shoes, and tie back long hair.
5. Wash your hands with soap and water immediately after completing this activity.

References

American Society for Engineering Education. 1987. *A national action agenda for engineering education.* Report of the ASEE Task Force on a National Action Agenda for Engineering Education. Washington, DC: American Society for Engineering Education.

NGSS Lead States. 2013. *Next Generation Science Standards: For states, by states.* Washington, DC: National Academies Press. *www.nextgenscience.org/next-generation-science-standards.*

National Academy of Engineering. 2005. *Educating the engineer of 2020: Adapting engineering education to the new century.* Washington, DC: National Academies Press.

National Academy of Engineering. 2008. *Changing the conversation: Messages for improving public understanding of engineering.* Washington, DC: National Academies Press.

National Academy of Engineering and National Research Council. 2009. *Engineering in K–12 education: Understanding the status and improving the prospects.* Washington, DC: National Academies Press.

National Research Council (NRC). 2012. *A framework for K–12 science education: Practices, crosscutting concepts, and core ideas.* Washington, DC: National Academies Press.

National Science Foundation. 2000. *Science and engineering indicators.* Biennial Report to Congress.

Pearson, G., and A. T. Young, eds. 2002. *Technically speaking: Why all Americans need to know more about technology.* Washington, DC: National Academies Press.

Portsmore, M. 2010. Exploring how experience with planning impacts first grade students' planning and solutions to engineering design problem. Unpublished Dissertation, Tufts University.

Websites

Engineering is Elementary: *www.eie.org*
Freedom Chair: *www.gogrit.us/lfc*
TeachEngineering: *www.teachengineering.org*

Book Resource

A Long Walk to Water; Park, L. S.; Age Range: 10–12; Lexile Level: 720L

Supporting Reading, Writing, and Discussion

Literacy instructional time is often precious in schools where reading, writing, and critical thinking skills have the highest priority. In the previous chapter, we talked about what engineering looks like for young children and how Novel Engineering can support students' engineering thinking and practice. Now, we will explore how Novel Engineering supports those high-priority English language arts (ELA) skills and can target key aspects of literacy learning, including purposeful use of text, perspective taking, close reading, intrinsic motivation, and student engagement. We'll show examples of students' Novel Engineering discussions to examine how the process can support reading comprehension and discussions about text. A well-designed Novel Engineering unit can provide students with the best of both worlds: engaging engineering experiences and rich literacy learning.

First, what are we talking about when we talk about literacy? Literacy is obviously a broad topic that encompasses many skills and concepts. For the purposes of this chapter, we will not focus on the key component skills that all readers and writers need to access and produce in order to read and write (e.g., word knowledge, decoding, spelling, written conventions). Instead, we will focus on how students make sense of what they read and how they use close reading of text to build understanding and engage in discussions. The concept of literacy can also encompass the writing process and the various forms of writing students need to thrive in school (e.g., notes, summaries, descriptions, arguments). Therefore, writing is part of our literacy focus; indeed, students write at numerous points throughout a Novel Engineering unit to communicate their ideas, document their work, and justify design decisions.

In the early stages of literacy instruction, students are learning how to read and transition to reading to learn. Reading with a purpose and writing become more intertwined. Students are expected to make more sophisticated inferences, use reading to prepare them for writing, and be able to write about what they have read. A variety of genres are included in classroom curricula, and students are expected to comprehend and respond to texts that include fiction, nonfiction, and historical narratives. Students are also expected to respond to what they have read through discussion and a variety of writing genres that include descriptive, expository, narrative, and persuasive.

We've found that students naturally connect with the people and animals in the books they read, whether they are real or fictional characters. Through our research, we've also seen Novel Engineering help teachers address core literacy goals in ways that allow for differentiation among students with different reading and writing abilities, yet maintain high levels of engagement across the board.

Dynamic Interactions Between Literacy Learning and Engineering

Novel Engineering leverages the alignment of literacy and engineering. Characters become the clients for whom students are engineering, so students pay close attention to the characters and settings as they find evidence to support their claims about what is happening in books and their engineering design choices.

If you think back to the *Wonder* example in Chapter 1, Samuel and Mateo based their decisions and design on the constraints of the character's situation and feelings. During their overall conversation about the design, Samuel and Mateo demonstrated an understanding of the text and its characters, and they communicated their ideas to each other while referring to specific character traits and aspects of the setting they gleaned from the text. In addition to these conversations, they completed a detailed planning document that outlined their ideas and explained their design as they presented it to the class. In other words, their teacher used the Novel Engineering project to support students' text discussions, writing, and oral presentation skills.

This dynamic relationship between the engineering design process (EDP) and students' interpretation of text creates a rich environment for literacy learning. During Novel Engineering units, we have seen students engage in a variety of desirable literacy practices, both independently and when prompted by a teacher. These literacy practices include the following examples:

- Taking the perspectives of the characters
- Considering relevant aspects of the physical setting of the narrative

- Referring to specific moments or descriptions in the text
- Grappling with unfamiliar concepts and vocabulary
- Taking notes, labeling diagrams, writing lists, or making drawings to support discussions
- Making substantive arguments using the text
- Writing narratives about the project as related to the text

As students work on these literacy skills, they also develop a set of engineering design skills. The engineering and literacy work are mutually beneficial. In other words, we see bidirectional benefits of combined engineering and literacy projects. Design can support reading comprehension, and the complex information and character perspectives in the book can create an authentic and engaging opportunity for engineering.

Other researchers have also found the integration of literacy instruction into other content areas is a way to support both literacy and content-area learning (Biancarosa and Snow 2004; Palincsar and Duke 2004).

- Integrated literacy and science instruction can have a greater positive effect on elementary school students' overall reading comprehension of science texts than instruction focused exclusively on reading (Guthrie et al. 2004; Guthrie, Wigfield, and VonSecker 2000; Morrow, Pressley, Smith, and Smith 1997).
- Widespread interest in interdisciplinary education at the elementary level has led to interventions supporting reading comprehension and content-area learning in subjects including science (e.g., Guthrie et al. 2004), mathematics (Halladay and Neumann 2012), and the arts (Grant, Hutchinson, Hornsby, and Brook 2008).

These bidirectional benefits are not only limited to reading. Student writing can also be enriched and supported during Novel Engineering projects through note taking, analyses of characters' needs and setting, written plans, and other forms of writing required to support the engineering process.

There are, in fact, many similarities between the writing process and the EDP. In both, students brainstorm and pick through ideas based on interest and then use evidence to support and build on an idea. Both processes are iterative and call on students to make changes based on self-evaluation and feedback from external sources. Also, both endeavors are done with a specific audience in mind and have a goal of sharing the final creation. When writing is used to support, document, and present the results of the engineering project, these two creative processes have the potential to reinforce and inform each other.

There are many skills that students need to use in both client-centered engineering and literacy. For example, interpreting or writing an analysis of a novel are certainly very different practices than designing a wheelchair, and they call on distinct skill sets and knowledge types. Both endeavors require students to consider diverse perspectives and settings. The text in a Novel Engineering unit is a resource for engineering that sets a context for messy human situations and relationships. In turn, engineering projects act as a platform and motivation to mine the text for relevant information.

Novel Engineering encourages close and purposeful reading of the text, frequently without prompting by the teacher. Students are routinely called on to apply their understanding of the text to inform their design choices. Therefore, students must infer the relevance and relative importance of information they glean from the text. While doing this, students are highly engaged in the text, often participating in spontaneous conversations about characters and the choices they make within the context of the text.

Providing Authentic Purposes

One of the reasons that integrated literacy and content-area instruction supports productive reading practices is because students engaged in these units are more likely to read and write for authentic purposes. In other words, reading comprehension is supported because students pay attention to relevant textual information so they can accomplish defined content-area goals. Purcell-Gates, Duke, and Martineau (2007) argue that authentic literacy occurs when students read and write text for purposes other than simply improving their reading and writing skills. In addition, authentic reading and writing serves sociocommunicative goals, such as reading to find specific information and writing to communicate information to an identified audience.

We can see students in Novel Engineering units use text in a purposeful way when they read for information about needs, resources, and relevant constraints. For example, in one fourth-grade classroom, during a unit on Jeanne DuPrau's *The City of Ember*, two students discuss what materials they should use in their design. Their conversation shifts as they think about what materials the characters would have at their disposal. In *The City of Ember*, the characters live in an underground city and are running out of resources. One of the students picks up the book and opens to a page that lists some of the resources the characters have.

She says, "You know how Doon's Dad um … has things like [*reads from book*] 'nails, pins, tacks, clips, springs, jar lids, doorknobs, bits of wire, shards of glass, chunks of wood,' and other small things that might be useful in some way?" At this moment, Sophie is engaging in a literacy practice that many teachers must often assign, coax, and reinforce. She is rereading the book to find evidence to support her conjecture. Sophie is also engaged in engineering as she considers what options the characters have for materials within the context of the book.

As Sophie spontaneously engages in a productive literacy practice, it enriches her recall and interpretation of the book. Research on authentic literacy generally supports the idea that reading is a functional tool and that reading skills may be supported when reading is done in the context of purposeful or authentic activities—rather than being taught in isolation (McGinley and Tierney 1989). Integrated units such as those in Novel Engineering provide a learning space in which reading serves a real purpose, rather than being the goal in and of itself.

Establishing a purpose for reading, either teacher-imposed or student-generated, has become a more common practice in reading instruction (McKeown, Beck, and Blake 2009). Students are more likely to work through a challenging text and engage with the text's concepts and information if they read with a conscious, engaging purpose. With Novel Engineering, students' purposes revolve around the client-centered engineering design they create as they read, write about, and discuss the text and its accompanying themes.

In Novel Engineering, students are challenged to design a prototype that could help characters facing a problem. Students are therefore encouraged to think about problems in the text, empathize with characters by considering their needs and preferences, and think about the constraints and resources described in the world of the text. One of the stated and genuine purposes for reading (and potentially rereading) the text, therefore, is to support the engineering project.

Engineers do not read novels to identify engineering problems, but they do interview clients, visit sites, and conduct other research that provides some of the same types of information available in a novel. The goal of Novel Engineering projects is not to directly replicate the experiences of a practicing engineer but to provide a relatively authentic and rich environment in which students can develop their understanding of the full EDP. Purcell-Gates, Duke, and Martineau (2007) suggest that authenticity can be described as a continuum, with activities being more or less authentic based on the nature of the task and how it relates to students' own questions and ideas. The activities in Novel Engineering units are therefore not perfectly authentic, but analysis of the text can serve as a substitute for direct interaction with a client and real-world settings.

Supporting Reading Comprehension

Understanding text is at the core of Novel Engineering units. In order to engage in client-centered engineering, students must examine the settings described in the text, as well as the needs, limitations, and preferences of characters. Peers and teachers can evaluate the success of a student's project in part based on the extent to which it makes sense for the characters and how well it could work within the world featured in the text.

As students move through the design process, therefore, they are called upon to read carefully, use text as evidence, think about the setting, make inferences about the characters, and explore how well various solutions might work. Their projects are informed by the text, but they also serve as an inspiration for engaging in productive literacy practices that support reading comprehension.

In addition, teachers can have a big impact on key aspects of the text that students are asked to consider when designing. Teachers can direct students to address specific themes or character traits, and they can help guide students' understanding of the text as they work to incorporate that understanding into their designs. Teachers should be prepared for students to notice aspects of the text and make interesting inferences that they (and we, as researchers) never saw coming.

To illustrate some of the ways that Novel Engineering projects can support reading comprehension, we are going to share an example from a fifth-grade class reading *From the Mixed-Up Files of Mrs. Basil E. Frankweiler* by E. L. Konigsburg. A key thing to note when reading about this presentation is that the teacher, on several occasions, focuses on a key vocabulary word taken from the novel: *inconspicuous*. In whole-class and small-group discussions, the teacher helps students explore what this concept means and why it matters to the characters, and directly teaches both the word and its definition.

At the end of the unit, students are asked to present their designs. A trio of girls has focused on the fact that the only money the two characters have after they run away from home is a large amount of loose change. Difficulties dealing with the change are described in the book, and these students infer other difficulties related to this source of funds, including the noise the coins could make and difficulties the characters may have hiding the money as they move around during the day. Therefore, the girls design a backpack that stores change. The goal of the backpack is to keep the coins quiet and prevent the sibling characters from standing out and being noticed.

During their class presentation, Ella and Laura use textual references to explain and justify their design decisions. For example, they refer to two details from the novel that describe how carrying lots of change around was "weighing [Jamie's] pants down" and "making noise." They also explain that they wanted to make something that would hide the money so "people wouldn't suspect." The girls are concerned that the design should help siblings Claudia and Jamie be *inconspicuous*, echoing the theme and vocabulary word their teacher emphasized, and many of the features of the backpack reflect that concern. During their presentation, they emphasize that they cut the change slots into the part of the money compartment that faces the wearer's back, keeping the slots hidden while Claudia or Jamie wear the backpack. Though they do not initially use the word *inconspicuous*, their design appears to embody their understanding of this important theme.

As the girls respond to questions from their classmates, they continue to explain their design choices in light of Claudia and Jamie's desire to be inconspicuous.

1. **Student 1:** Won't that still make noise—the change?

2. **Ella:** There's padding.

3. **Student 2:** Would people be able to see from the top?

4. **Laura:** No, because there's going to be a cover here. [*gestures to the top of the backpack*]

5. **Student 3:** I have a question. Why did you cover it with fabric?

6. **Ella:** Oh, because if it was just cardboard, people would notice that cardboard was on their back and they would be suspicious.

7. **Teacher:** What's the word?

8. **Ella:** *Inconspicuous.*

9. **Teacher:** Good, so they want to be inconspicuous.

Ella and Laura's classmates pick up the theme of staying inconspicuous and evaluate the design using this criterion. Kate is concerned that the change will make noise, presumably calling attention to the protagonists. Eric notices that they have not attached the top flap yet and thinks about whether people would be able to look inside.

Ella then explains a design decision—covering the prototype with fabric—saying that a cardboard backpack would make people "suspicious." At this point, the teacher prompts Ella to use the word *inconspicuous*, helping her link the term from the book with the criterion the class is using to evaluate the design. The girls again emphasize the need to avoid notice when they explain why they chose to make a backpack instead of a purse. Specifically, they want both protagonists to use it and thought it might draw attention if Jamie (a boy) carried a purse. In this case, the presentation also helped the class as they grappled with understanding new and important vocabulary.

Novel Engineering units obviously help students explore the characters, settings, and themes of novels, but which books work best for these sorts of projects? One challenge for designing integrated literacy and content-area units is the selection of appropriate texts. It is important that the books used in Novel Engineering units are accessible and interesting to students. The use of high-quality, engaging texts is essential for both developing reading comprehension skills and learning relevant science and engineering concepts (Guthrie et al. 2004; Hapgood, Magnusson, and Palincsar 2004; Morrow et al. 1997).

Using high-quality expository and narrative texts in integrated instruction supports content-area and literacy learning, but it also provides the reader with opportunities for an emotional response, personal association, evaluation, and insight beyond dry, fact-based texts (Morrow et al. 1997). Although the books aren't necessarily specific to STEM disciplines, students get the chance to engage with these disciplines as they design. In Novel Engineering, a wide range of books can be used, so teachers have the freedom to pick books that address student goals, are within students' targeted reading level, and involve topics that are interesting and engaging.

As always, the quality of reading instruction is still a crucial component of literacy learning in integrated literacy and content-area instruction. Students will not develop productive reading comprehension strategies or improve their writing simply because they are engaged in hands-on activities or are focused on a particular theme. Literacy learning will always depend on skilled teachers who rely on tried-and-true best practices for their instruction. Practices for supporting reading comprehension include teaching relevant vocabulary, activating background knowledge, helping students generate text-dependent questions, summarizing, and supporting comprehension monitoring while reading. Supporting writing development, in turn, still involves methods of generating and keeping track of ideas, note taking, organizing text, and revising.

Depending on students' grade level and specific needs, teachers may also rely on reading to students, conducting daily written responses to or reflections on text, and supporting students as they engage in retelling, rewriting, discussing, keeping track of books read, and following key themes within and across books.

Perspective Taking and Empathy

Perspective taking and empathy are key activities that help students interpret text, understand characters, and make inferences. These reading comprehension skills also connect strongly with engineering design. In the following exchanges, we see students engage in both activities. Kyle and Sally are designing a solution to help Ralph, the mouse who is the main character from Beverly Cleary's *The Mouse and the Motorcycle*. Ralph, unfortunately, has found himself trapped in a trash can with his small but functional motorcycle. Kyle and Sally are trying to design a tool he can use to escape. Although the character is a fictional mouse, the students have set aside the ridiculousness of the situation and put themselves in Ralph's place as they discuss the design.

1. **Kyle:**   I think we need to have it a little smaller, actually. If [Ralph] needs to, like, hold it, that would be, like, too heavy.

2. **Sally:**   Maybe [Ralph] could get string, and he has claws so he can grab it.

Kyle and Sally are thinking about the mechanics of the tool, but they are also considering their client, Ralph, and what size tool he would need to make it functional. They know that some things are too heavy for a mouse but also that a mouse has resources, namely claws, that could help him use the mechanism they are working to design.

Students can also easily identify with the challenges faced by young characters. In another classroom, students are reading *Clementine* by Sara Pennypacker. The teacher leads a classroom discussion in which students discuss designs that could help a character named Margaret, whose hair was cut off by her best friend. Margaret is very proud of her hair and is embarrassed by what it looks like after it is cut. Two students, Holly and Mary, focus on Margaret's potential embarrassment about being seen in public—especially at school. They think a wig might be a possible solution, but they also talk about the potential downside associated with that option.

1. **Holly:**   Um, me and Mary were thinking of pros and cons too and we—

2. **Teacher:**   Great. For the wig?

3. **Holly:**   Huh?

4. **Teacher:**   For the wig idea?

5. **Holly:**   Yeah for the wig, and one of the pros was that she could still go to school and no one would really notice it, but the bad one would be, what happens if, like, she goes upside down or something or—

6. **Mary:**   On monkey bars.

The girls feel that a wig would make Margaret feel less embarrassed, but they also point out that if she is playing and "goes upside down," her wig may fall off, which would be a further source of embarrassment. Whether students are thinking of a mouse or a short-haired girl, they are using information they read in a text to empathize with characters and guide their engineering brainstorming.

Client-centered engineering requires a focus on clients' needs, preferences, and abilities. Whether a client has a unique physical ability (e.g., grabbing things with claws) or likes to engage in an activity that may find her upside down, it is important to factor this understanding into a design. Designs may succeed or fail based on a designer's ability to think about real clients who will use the product in a range of settings.

Integrated units also support literacy learning in part because students must consider characters' perspectives and abilities in order to create a design that would function in concrete, visible, and meaningful contexts. With this in mind, many researchers promote the use of extended projects and hands-on activities during integrated units (Guthrie et al. 2004; Hapgood, Magnusson, and Palincsar 2004; Romance and Vitale 1992). These instructional approaches prompt students to apply text-based knowledge and potentially require them to recall and interpret relevant material from the text. In other words, the concrete process of building something with their hands helps students with the more abstract processes of adopting new perspectives and empathizing with characters' needs.

After the class reading *Clementine* finishes a whole-class discussion, partners complete a worksheet that asks them to pick one of the ideas the class brainstormed and explain why they think it is the best solution (see Figure 3.1). In this document, as in the class discussion, students are bringing their previous knowledge about the comfort of wigs and that young students play on monkey bars at school. This knowledge frames their thinking about the suitability of the wig solution, and their answers show an awareness of Margaret's feelings and an understanding of the "cons" they need to think about as they plan a wig for Margaret.

This worksheet (and students' related work) meets the teacher's ELA and engineering goals. Because students are moving toward building a real-life prototype of their solution, they are immersed in the world of the text and the practical realities of design, therefore taking full advantage of the benefits of integrated, hands-on, project-based learning.

Addressing Multifaceted Literacy Goals

Teachers have found that Novel Engineering helps them address multiple goals they have for their students within engineering and literacy—both teacher-derived goals and mandated standards. Table 3.1 (p. 46) highlights which *Common Core State Standards* (CCSS) can be addressed at certain points along the Novel Engineering trajectory. (See Appendix C, p. 226, for a more detailed description of *CCSS* met by Novel Engineering.) Just as with the engineering aspects of Novel Engineering, students will not be able to address all the standards in one unit. The information in this table can be used as a resource as teachers decide on how they want to structure Novel Engineering units and when they want students to focus on specific literacy curriculum standards.

Figure 3.1: Student work from *Clementine*

In the book, *Clementine*, by Sara Pennypacker, Margaret is faced with a problem: She and Clementine have cut off all Margaret's hair!

The solution I think that would work best for Margaret right now is:

the wig

Three reasons I think this is the best solution for Margaret

are:

1. Margret when she goes to the bathroom when she takes it off it's like a broke for her.

2. Know One will notice that she has a wig on.

3. She can still go to school!

Additional information/thoughts:

Might fall off. feels fake. ~~Might~~ Might not be her hair color.

Meeting the Needs of Students With Different Skills and Learning Styles

Novel Engineering allows teachers to differentiate instruction for students and meet individual needs even as students are working toward the same goals. As we know, students within a single classroom have different reading and writing levels but still need access to the same texts and concepts. Many students are expected to develop comprehension, composition, and discussion skills while they are still developing component reading and writing skills (e.g., word reading, spelling, grammar, vocabulary). The teacher must also keep students engaged

Table 3.1: *Common Core State Standards* for English language arts

Common Core State Standards for English Language Arts and Literacy	Novel Engineering Arc					
	Read Book and Identify Problems	Solve Problems and Brainstorm Solutions	Design Solutions	Get Feedback	Improve Solutions	Reflect and Share
Reading						
CCRA.R.1 Read closely to determine what the text says explicitly and to make logical inferences from it; cite specific textual evidence when writing of speaking to support conclusions drawn from the text.	X					X
CCRA.R.2 Determine central ideas or themes of a text and analyze their development; summarize the key supporting details and ideas.	X			X		X
CCRA.R.3 Analyze how and why individuals, events, or ideas develop and interact over the course of a text.	X	X				X
Writing						
CCRA.W.1 Write arguments to support claims in an analysis of substantive topics or texts using valid reasoning and relevant sufficient evidence.						X
CCRA.W.2 Write informative/explanatory texts to examine and convey comples ideas and information clearly and accurately through the effective selection, organization, and analysis of content.						X
CCRA.W.3 Write narratives to develop real or imagined experiences or events using effective technique, well-chosen details and well-structured event sequences.						X
CCRA.W.6 Use technology, including the Internet, to produce and publish writing and to interact and collaborate with others.			X	X		X
CCRA.W.9 Draw evidence from literary or informational texts to support analysis, reflection, and research.			X			X

(continued)

Table 3.1: *Common Core State Standards for English language arts (continued)*

Common Core State Standards for English Language Arts and Literacy		Novel Engineering Arc					
		Read Book and Identify Problems	Solve Problems and Brainstorm Solutions	Design Solutions	Get Feedback	Improve Solutions	Reflect and Share
Writing *(continued)*	CCRA.W.10 Write routinely for over extended time frames (time for research, reflection, and revision) and shorter time frames (a single sitting or a day or two) for a range of tasks, purposes, and audiences.			X	X		
Speaking and Listening	CCRA.SL.1 Prepare for and participate effectively in a range of conversations and collaborations with diverse partners, building on others' ideas and expressing their own clearly and persuasively.		X	X	X	X	
	CCRA.SL.2 Integrate and evaluate information presented in diverse media and formats, including visually, quantitatively, and orally.			X			
	CCRA.SL.3 Evaluate a speaker's point of view, reasoning, and use of evidence and rhetoric.			X	X	X	
	CCRA.SL.4 Present information, findings, and supporting evidence such that listeners can follow the line of reasoning and the organization, development, and style are appropriate to task, purpose, audience.		X	X	X	X	
	CCRA.SL.5 Make strategic use of digital media and visual displats of data to express information and enhance understanding of presentations.						X
	CCRA.SL.6 Adapt speech to a variety of contexts and communicative tasks, demonstrating command of formal English when indicated or appropriate.		X	X	X		

and motivated when the text becomes more difficult and the vocabulary becomes more complex. As any teacher knows, this is especially important with students who may be frustrated by their own reading skills and have a harder time engaging with literacy in the classroom.

We have seen Novel Engineering work well for students with a variety of learning profiles, including those who have special accommodations to address their individual needs. We have worked with students in public school general education classrooms and students with learning-based language disabilities in private school programs. When we look at students' engagement and final engineering solutions, we see no difference in their conversations or the functionality of their designs. In fact, students who often feel unsuccessful in literacy-based activities are motivated to participate in both aspects of Novel Engineering units, feel positive about their experiences, and are proud to share their ideas.

Literacy and the Novel Engineering Trajectory

One goal of Novel Engineering is to use a project-based learning approach in content-area instruction. In Novel Engineering units, students work in pairs or small groups, read interesting texts, choose the problem and design solution, and then build a prototype—all elements shown to support content-area and literacy learning in integrated science and literacy units (Guthrie et al. 2004). Students can read a book as part of a full-class or small-group interactive read-aloud or during independent reading time. The promotion of note-taking strategies and the discussion of key events and vocabulary happens as part of the reading process. Teachers also set a purpose for reading, letting students know they should attend to the problems characters face and the resources in their environments.

In the following sections, we share examples of student work related to a few of the steps in the Novel Engineering trajectory (see Figure 3.2)—scoping problems, designing solutions, testing the solutions, getting feedback, and sharing culminating activities.

Scoping Problems and Designing a Solution

In this section, we look at the literacy embedded within students' early work in scoping problems and designing a solution. We see students, with support from teacher-designed worksheets, engage in taking notes, labeling diagrams, writing lists, and making drawings to support discussions. In a fourth-grade classroom reading Roald Dahl's *James and the Giant Peach*, a reading group composed of four students identify a problem from a scene in the book (when James and his friends are flying through the air on a magically enlarged peach and cloud people begin throwing hailstones at them).

Figure 3.2: The Novel Engineering trajectory

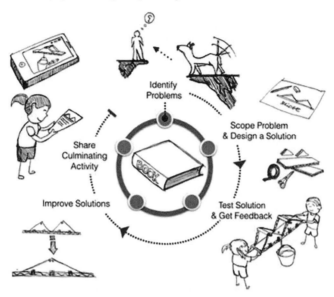

As we look in on Manny, Aidan, Stuart, and Jack, we see that they are engaged in desirable literacy practices as they have a collaborative discussion, build on one another's ideas, and contribute to the overall discussion. The teacher has provided a planning document (see Figure 3.3, p. 50), and the students have just started to address the first questions on the sheet: What problem did your group choose? Why is this a problem?

1. **Aidan:** [*reads worksheet*] What is the problem ...

2. **Jack:** Isn't it obvious?

3. **Aidan:** [The hailstones are] going super-fast, so if they get hit, they're probably going to fall and get hurt.

4. **Jack:** Or they're going to fall from 1,000 feet in air, hit the water, and probably destroy all of their bones.

The next question on the worksheet is "How did the characters deal with the problem?" This question, in contrast to the first one, does not have a clear answer that can be elicited from the text. Perhaps because of this ambiguity (or because the visiting researcher misremembered the book), the boys pause to summarize this scene and emphasize key details from the story. As they fill out the worksheet, however, they make some nuanced observations and inferences about the scene and again refer to specific details from the text. A researcher observes this part of the process and interacts with the students.

Figure 3.3: Student work from *James and the Giant Peach*

> What problem did your group chose? Explain why.
> My group chose the problem when the cloud-men were throwing hailstones at the group. We chose this problem because we knew we could come up with a great solution to solve the problem that would work.

> Why is this a problem? Explain.
> It is a problem because the hailstones are being thrown very fast and if they get hit they will get hurt and fall into the water.

1.	**Stuart:**	[*reads worksheet*] How did the characters deal with the problem? Was the problem solved in the book? Explain how. What should we write for this one? The problem wasn't solved.
2.	**Jack:**	It's not exactly solved.
3.	**Researcher:**	Well, what happens in the book?
4.	**Jack:**	Well, they just float away. The cloud men keep throwing at them; they just don't get them.
5.	**Researcher:**	Yeah, they just ... run away.
6.	**Jack:**	Well, they can't run away because they're floating on a giant peach ... propelled by ...
7.	**Researcher:**	Wind.
8.	**Stuart:**	No, um, seagulls.
9.	**Jack:**	Not wind. Seagulls.
10.	**Researcher:**	Oh yes, that's what it was. I can't remember.
11.	**Jack:**	Like 500, uh ...
12.	**Manny:**	Five-hundred-and-two. [*references the page describing the number of seagulls*]

Both Jack and Stuart agree that the book's characters did not devise a specific solution to their plight. Jack then notes that the characters waited until they were out of range. He also notes that, technically, they did not run away because they were stuck on a giant floating peach. When the researcher suggests that the peach was propelled by wind, Jack, Stuart, and Manny all respond by noting that the book portrays the peach as being pulled through the air by seagulls. Manny even notes the exact number of gulls from the book.

Apparently undeterred by the unrealistic scenario, the students initially plan to design a glove the characters could use to catch the hailstones—similar to one student's baseball glove. Manny, however, raises a concern.

1.	**Manny:**	Yeah, but how? Wouldn't the cloud men just aim at your arm?
2.	**Stuart:**	Maybe we could move our arm around.
3.	**Jack:**	Yeah. If it was just your arm, it would probably break your hand.
4.	**Manny:**	Yeah.
5.	**Jack:**	It said they were throwing it really, really hard. Like bullets.
6.	**Manny:**	Enough to break your arm.
7.	**Aidan:**	Yeah, it said they were throwing it in the sky and that they were throwing them as fast as they could ... like bullets.
8.	**Stuart:**	Oh! So we could, um, design something ... like a trap, so it's bigger. It's like, if we could attach sticks to it, put elastics to both ends so when they throw it, it just, it just goes back into the trap and then goes "zoom!" [*gestures up and away*]
9.	**Aidan:**	Maybe it could be like a tube and [the hailstone] goes back. Like a tube, and it has a couple elastics at the back end and when it goes in, [the hailstone] stops and shoots back out.
10.	**Jack:**	If you put [the hailstone] back in, then it would stay in there, and it wouldn't be like when they threw it, you caught [it].
11.	**Manny:**	You'd need bulletproof materials.

Because they are focused on building something that will work in the world of the book, the students use information from the text to discuss concerns, establish constraints, and consider alterations to their design idea. They also use relevant, specific information from the book and references to the text, such as the speed of the hailstones being thrown, without direct prompting from the teacher or researcher. The information in the book helps them establish criteria for a potentially successful design.

In addition, the groups makes inferences about how the hailstones would affect a person's arm, consider how the design would or would not address this issue, and discusses the materials that might be needed for their design. Ultimately, the students are also constrained by the realities of their classroom, including time limits and a narrow range of available materials. Bulletproof materials are not available for their classroom project, but their attention to design features, material properties, and functionality are at this point strongly influenced by contextual elements.

The worksheet and interaction with the researcher prompt the students to reconsider events in the book through an engineering lens and to refer to key details that support the points they are making. Perhaps because these students spend a good deal of time talking about the problem faced by James and his friends, they are quick to generate two responses to the question about the problem they are addressing. In the first case, Aidan refers to the speed with which the cloud men throw hailstones. In the second, Jack refers to the fact that James and his friends are portrayed as floating high above the ground—high enough to be in the clouds. Jack also infers that falling from this distance would result in a catastrophic injury for a human. Though the answer to the worksheet question seems obvious to them, it nevertheless provides them with an opportunity to articulate and leave a record of their thinking.

In the following snippet, taken from a different group of students in the same classroom, a researcher asks a pair of boys about their plans. The students are drawing a design and trying to figure out what they could build that would lift the giant peach, and are considering a lever system.

1. **Researcher:** Wait, can you tell me what you're drawing here?

2. **Charles:** We're trying to figure out which lever they would like—a lever to go up to pull the peach up with rope or something heavier to put it on to lift up the peach.

3. **Mark:** Let's draw the design for it.

4. **Charles:** I was thinking to cut cardboard here [*draws*] and then do another cardboard right over here.

5. **Mark:** Well, my idea was this—so let's just draw our ideas for that. (See Figure 3.4.)

In this moment, the students' drawings act as tools for their discussion. Each group of students uses one piece of paper to communicate, working on the same drawing (see Figure 3.5). The drawings help students refine their own ideas and communicate with each other.

Figure 3.4: Sketch created by Mark to show his idea

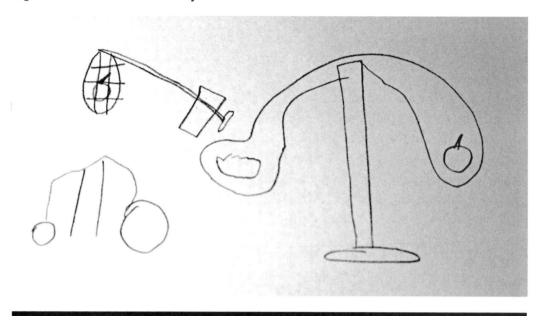

Figure 3.5: Students using a planning drawing to communicate ideas

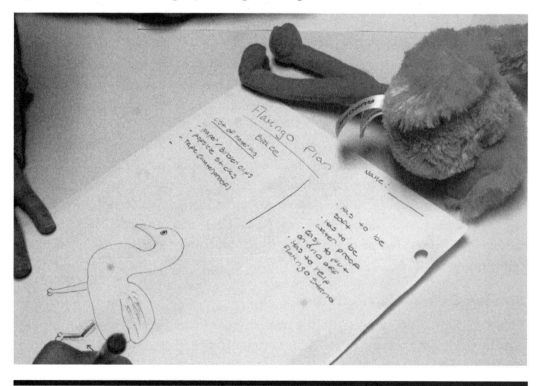

In these interactions, the book offers key resources for the engineering tasks and students use the text to justify their own design decisions. Even as the projects are influenced by their interpretations of the text, there is evidence from these interactions that doing the engineering project prompts students to focus on key elements of the text. Students engage with the text as they plan and evaluate their ideas.

In addition, their interaction with the text is productive because they engage in behaviors that are emphasized in *CCSS* for English language arts. They discuss and collaborate; refer to details in the text when making inferences and clarifying ideas; focus on a particular theme of the novel; and present their ideas using appropriate facts and relevant, descriptive details. Therefore, given a purposeful, engaging task and learning environment, students practice valuable literacy behaviors with minimal additional prompting by their teacher.

Testing the Solution and Getting Feedback

Another aspect of engineering and Novel Engineering is that designers need to make changes based on feedback that comes in the form of physical tests and peer feedback. We'll look at one third-grade classroom that is reading the book *If You Lived in Colonial Times* by Ann McGovern as part of their social studies curriculum. In response to the book, a group of students designs and tests a water filter prototype they hope will keep the colonial settlers from getting sick from drinking dirty water.

The students have built an initial prototype and, as they test it, one of them begins to think about the types of materials colonists might have had available to them. Since the teacher didn't specify if students were bound by colonial-era restraints in selecting their materials, it was left up to students to decide for themselves, within reason.

"Wait, uh, guys, can we step back a second? They didn't have cotton balls in colonial times." This revelation causes the group to look at the materials they are using in an attempt to remain true to what they know about the time period. Ultimately, they decide that cotton balls are an acceptable stand-in for wool, which they assume colonists likely had. As they test and refine their prototype, they keep a written record of their test results and refer back to it as they are redesigning.

Sharing Culminating Activities

As we've shown, interaction with the text happens throughout the design process in Novel Engineering, but a culminating activity gives students a time and place to reflect on their process and gives teachers another chance to assess students' understanding of the text. There are a variety of culminating activities that meet these criteria. Examples include writing an analytic essay, preparing a comic strip such as the one in Figure 3.6 for *Tuck Everlasting*, making a poster (Figures 3.7a and 3.7b, p. 56), or creating an oral presentation. In Chapter 13, we discuss different types of culminating activities in more detail and give examples of student work. Culminating activities can be presented in written or oral form. As you pick tasks for your students, keep in mind that the tasks should not only help you assess your students' understanding of the text but also address writing and speaking goals.

One type of writing activity involves asking students to write a chapter that would be in the book if characters had access to student designs (see Figure 3.8, p. 57). For example, one student wrote about a "lock picker," designed to get a character out of jail and started the chapter: "I'm in jail. It's dark and lonely, but

Figure 3.6: Comic strip done by students as a culminating activity for *Tuck Everlasting*

Figure 3.7a: Advertisement done by students as a culminating activity for *Tuck Everlasting*

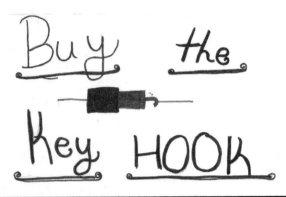

Figure 3.7b: Advertisement done by students as a culminating activity for *Tuck Everlasting*

Figure 3.8: "Missing" chapter written by students as a culminating activity for *Tuck Everlasting*

Chapter X

The stranger walked through the wood with poor Jesse, who had a bleeding leg, in his arm.

"Tell me where the water is fool."

"That... house... there," managed to say Jesse.

A huge house stood stickin between two trees. It looked like a regular log cabin, only different somehow.

"You hid the spring in a house!" added the stranger.

He dropped Jesse and ran into the house. Before Jesse could warn him, the stranger fell into a hole.

"AAHH!" Bellowed the stranger as he plunged into the darkness of the hole.

After a few seconds, oh it was ran out, the hole, the stranger hit the ground. Suddenly the stranger bellowed, again.

"AAHH!" "Poisoness Snakes!" yelled the stranger. "I will get for this Tooks!!"

He started cursing like crazy, but was cut off with death.

After a few minutes, Jesse got up and walked home.

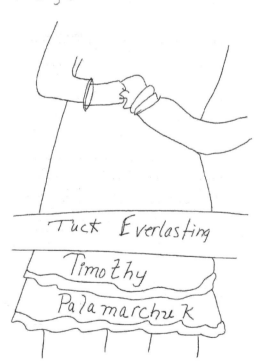

Tuck Everlasting

Timothy Palamarchuk

what the [constable] doesn't know is that I remembered to bring my special lock picker." The chapter goes on to describe how the device works and how it would affect the end of the book. To write this, the student needs to have a strong understanding of the trajectory of the story and how different characters might react to new situations.

Wrap-Up

Novel Engineering addresses literacy goals as students move through the design trajectory, providing a context for literacy that engages students and gives them authentic purpose for reading and designing. Novel Engineering units offer opportunities and motivation for students to read texts thoroughly, interact with the content, and identify with the characters as they help them solve problems. Evidence of this is apparent in students' conversations, written work, and engineering designs.

Safety Notes

1. Wear safety goggles or glasses with side shields during the setup, hands-on, and takedown segments of the activity.
2. Use caution when using hand tools that can cut or puncture skin.
3. Use only GFI-protected circuits when using electrical equipment, and keep away from water sources to prevent shock.
4. Secure loose clothing, remove loose jewelry, wear closed-toe shoes, and tie back long hair.
5. Wash your hands with soap and water immediately after completing this activity.

References

Biancarosa, G., and C. E. Snow. 2004. *Reading next: A vision for action and research in middle and high school literacy*. A report to Carnegie Corporation of New York. Washington, DC: Alliance for Excellent Education.

Grant, A., K. Hutchinson, D. Hornsby, and S. Brooke. 2008. Creative pedagogies: "Artfull" reading and writing. *English Teaching: Practice and Critique* 7 (1): 57–72.

Guthrie, J. T., A. Wigfield, P. Barbosa, K. C. Perencevich, A. Taboada, M. H. Davis, et al. 2004. Increasing reading comprehension and engagement through concept-oriented reading instruction. *Journal of Educational Psychology* 96 (3): 403–423.

Guthrie, J. T., A. Wigfield, and C. VonSecker. 2000. Effects of integrated instruction on motivation and strategy use in reading. *Journal of Educational Psychology* 92 (2): 331–341.

Halladay, J. L., and M. D. Neumann. 2012. Connecting reading and mathematical strategies *The Reading Teacher* 65 (7): 471–476.

Hapgood, S., S. J. Magnusson, and A. S. Palincsar. 2004. Teacher, text, and experience: A case of young children's scientific inquiry. *Journal of the Learning Sciences* 13 (4): 455–505.

McGinley, W., and R. J. Tierney. 1989. Traversing the topical landscape: Reading and writing as ways of knowing. *Written Communication* 6 (3): 243–269.

McKeown, M. G., I. L. Beck, and R. G. K. Blake. 2009. Rethinking reading comprehension instruction: A comparison of instruction for strategies and content approaches. *Reading Research Quarterly* 44 (3): 218–253.

Morrow, L. M., M. Pressley, J. K. Smith, and M. Smith. 1997. The effect of a literature-based program integrated into literacy and science instruction with children from diverse backgrounds. *Reading Research Quarterly* 32 (1): 54–76.

Palincsar, A. S., and N. K. Duke. 2004. The role of text and text-reader interactions in young children's reading development and achievement. *Elementary School Journal* 105 (2): 183–197.

Purcell-Gates, V., N. K. Duke, and J. Martineau. 2007. Learning to read and write genre-specific text: Roles of authentic experience and explicit teaching. *Reading Research Quarterly* 42 (1): 8–45.

Romance, N. R., and M. R. Vitale. 1992. A curriculum strategy that expands time for in-depth elementary science instruction by using science-based reading strategies: Effects of a year-long study in grade four. *Journal of Research in Science Teaching* 29 (6): 545–554.

Book Resources

Clementine; Pennypacker, S.; Age Range: 6–9; Lexile Level: 790L

The City of Ember; DuPrau, J.; Age Range: 5–12; Lexile Level: GN520L

From the Mixed-Up Files of Mrs. Basil E. Frankweiler; Konigsburg, E. L.; Age Range: 8–12; Lexile Level: 700L

If You Lived in Colonial Times; McGovern, A.; Age Range: 7–10; Lexile Level: 590L

James and the Giant Peach; Dahl, R.; Age Range: 7–10; Lexile Level: 870L

The Mouse and the Motorcycle; Cleary, B.; Age Range: 7–10; Lexile Level: 860L

Tuck Everlasting; Babbitt, N.; Age Range: 9–11; Lexile Level: 770L

Wonder; Palacio, R. J.; Age Range: 8–12; Lexile Level: 790L

Section II

Case Studies From the Classroom

Recognizing Children's Productive Beginnings

> As you read this chapter, think about what you would consider to be the beginnings of engineering in students' discussions and work. How do you see engineering and literacy supporting each other in this unit? How would you respond to students in different moments?

For our first case study, we visit a fourth-grade classroom reading *From the Mixed-Up Files of Mrs. Basil E. Frankweiler* by E. L. Koningburg. The book is about two runaways, Claudia and Jamie, who stay in a museum. The purpose of this chapter is to show what the beginnings of engineering look like in a fourth-grade classroom and to provide insight from a teacher's first experience with Novel Engineering. At several key moments, the teacher, Maggie, discusses her experiences planning and implementing her first Novel Engineering project, and she reflects on her students' discussions and ideas as the project unfolds. She provides a teacher's perspective on planning a Novel Engineering unit and making instructional decisions when responding to students' questions, conversations, and ideas about the book and their engineering designs.

Productive Beginnings

As children learn to do engineering, we often talk about looking at their initial activities for the productive beginnings of engineering. Productive beginnings show students doing aspects of engineering, maybe not as dexterously as professional engineers, but with some of the essential elements (e.g., patterns of thinking and behaviors). By focusing first on students' skills and productive behaviors, we are better able to help them strengthen those skills and abilities in engineering. This approach stands in contrast to one that might focus or provide critiques on what students are doing poorly or differently. This chapter shows that students can easily engage with many aspects of engineering as beginners, and it discusses how we can support them as they gain expertise.

Maggie set up a Novel Engineering task to be open-ended, providing students with expectations and learning goals while allowing them to follow their own ideas and interests. Ultimately, her students were able to take a messy, ill-defined problem and navigate their way through the engineering design process (EDP) to arrive at a functional solution. They were able to use information from a novel to define constraints and criteria that would affect their designs; indeed, their use of the text was evident from their conversations while designing and building and from their presentations. In addition, their design iterations reflected measured decisions based on feedback from tests, adults, and classmates.

We will focus on two groups of students who came up with markedly distinct engineering solutions. Both groups created functional prototypes that uniquely solved the problem. Moreover, in their design processes, both groups considered implicit and explicit constraints while also using evidence from the text. Before talking specifically about these two groups, we'll give more information about how Maggie set up and supported the Novel Engineering unit in her classroom.

Maggie's fourth-grade class of 22 students was in a suburban school district outside of Boston, Massachusetts. She chose *From the Mixed-Up Files of Mrs. Basil E. Frankweiler* by E. L. Konigsburg because it was already part of her English language arts (ELA) curriculum, the characters were accessible to students, and there were several real-world challenges in the book for which Maggie believed students could engineer solutions.

> **Although many different texts can be used for Novel Engineering units, you should pick one for which you can anticipate the problems and solutions students may identify. Are there any books you currently teach that would meet the criteria Maggie used?**

As we describe the classroom's trajectory with her Novel Engineering task, we focus on how Maggie set up individual tasks within the larger activity and students' engagement with these tasks (Table 4.1). Maggie uses the text as a

Table 4.1: Trajectory of Novel Engineering unit in Maggie's classroom

Session	Activity
Days 1–4: Reading	Teacher reads aloud with full-class discussions.
Day 5: a. Answering "What Is an Engineer?" b. Identifying Problems c. Introducing the EDP	• Teacher introduces engineering and the task. • Students identify problems the characters in the book are having and talk about problems that could and could not be solved with engineering. • Teacher introduces the EDP.
Day 6: Planning	Students plan, research, and design; this can include discussing ideas, sketching ideas, referring to the book to support ideas and consider constraints, and considering materials for the design.
Day 7: a. Receiving Materials b. Building	Students receive materials and begin to build.
Day 8: a. Building b. Testing c. Presenting	Students build and test; some students finish building and present their designs.
Day 9: Presenting	Groups give presentations.
Day 10: Summarizing	Class discusses and summarizes the book.
Day 11: a. Completing Writing Assignment b. Assessing Students	• Students write alternative endings to the book that include their solutions. • Students' writing assignment and presentation content can be used for assessment.

read-aloud, pausing as she reads over several days to spark class discussions. Beyond reading the book together, Maggie's class completes the building and writing tasks over seven sessions lasting 45–60 minutes each. We'd like to stress that this is just one model. We've seen a range of time allotted for building, from as little as two hours with older students to multiple sessions over days or weeks.

> **Maggie's Reflection**
>
> *My goals in implementing this first Novel Engineering unit were twofold: I wanted to provide my students with an authentic engineering experience as well as engage them in important literacy practices, such as making sense of the text and drawing on evidence for their inferences. I recognized that to achieve these goals I'd have to let go of my expectations for students and let them develop their own ideas. I deliberately planned how to take a more hands-off approach to let students take the lead in scoping problems, brainstorming solutions, and realizing their designs. I recognized that relinquishing this control could be a bit intimidating and uncomfortable, but I thought the payoff for my students' learning would be worth it.*

What Is an Engineer?

After reading Chapter 6 of *From the Mixed-Up Files of Mrs. Basil E. Frankweiler* aloud to students, Maggie decides to have a class discussion to introduce engineering. This begins with students breaking into pairs to discuss the question "What is an engineer?" Next, the pairs share with the class what they think engineers do. Responses include "An engineer is someone who builds and designs things using materials," "They fix things," and "They use math and science."

1. **Student 1:** Um, an engineer is someone who has materials [and] not makes them but builds stuff with them and designs it.

2. **Maggie:** Do they have to build?

3. **Student 1:** Yes.

4. **Student 2:** No.

5. **Student 1:** Not exactly.

6. **Maggie:** Hmm, all right. Let's think about that.

7. **Student 3:** Um, like, an engineer can take bugs out of computer chips.

8. **Maggie:** Okay, so, technology. What else do you have to do in order to take something out or to fix something? What do you have to do in order to do that? We do it in math a lot.

9. **Student 3:** Problem solve.

The class then talks about different kinds of engineering and what they think engineers do, which leads to a fuller discussion of what engineers do when they are working. Maggie is responsive to students' ideas and comments and allows the class discussion to follow those ideas. As a group, students come to a shared understanding of what engineers do: "An engineer is a problem solver that uses math and science to create or construct things and helps solve problems for people by making or improving things."

Maggie then shifts the discussion to encourage students to reflect on their own abilities. She asks them if *they* have ever been engineers. Many students start to make connections between their own problem solving and the problem solving that engineers do (Figure 4.1).

Next, Maggie brings the conversation back to the book: "We will be reading the book as engineers. What are the problems the characters encounter that an engineer could maybe help with? Let's start a list of problems that we might,

Figure 4.1: Documentation of students' ideas about engineering

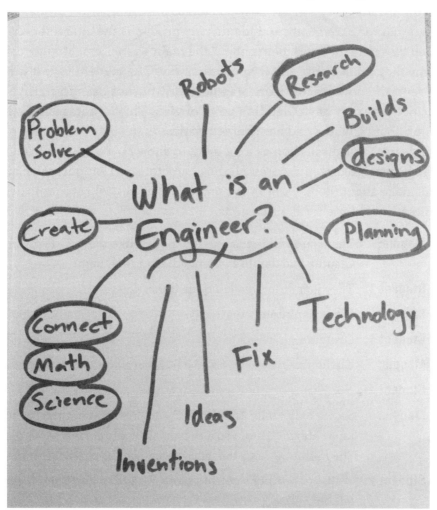

or engineers might, be able to help Claudia and Jamie with specifically. Other characters might have some problems that engineers could help us solve, so what we're going to do today is give a little summary of what we've read so far, and if there are any problems, we can start writing them down. As we read each day, we'll come up [*points to a large sheet of paper attached to the board*] with any problems and start listing them on the board."

> **Depending on their age, students may need help understanding what a problem is or what types of problems engineers can solve. How might you introduce engineering to your students? Chapter 10 (p. 163) discusses ways to do just that.**

Maggie's students spend the next portion of the class period discussing what happened thus far in the book and identifying problems the characters faced. Prompted by a student's comments about the main characters, Maggie asks students why they think Jamie and Claudia ran away. One student says it is because their parents did not care for them. Maggie gently pushes back to clarify this notion since they only have Claudia's point of view. This initiates a more in-depth discussion about Claudia and her characteristics.

Maggie uses this discussion as a moment to show how evidence from the book can be used to better understand the characters and support the claims students make about them. Later in the process, this will help students define constraints and criteria for their designs.

1. **Maggie:**   That sounds like the parents did not like them. Do you think that Claudia and Jamie's parents did not like them?

2. **Student 1:** They just didn't really notice them.

3. **Maggie:**   Whose opinion is that?

4. **Student 1:** Claudia's.

5. **Maggie:**   Claudia's. Okay, do we know how her parents really treated her?

6. **Student 1:** No.

7. **Maggie:**   So, we only really know how Claudia feels about this. What do we know about Claudia as a person? Is she a normal sixth-grade girl? She might be. Let's talk about some of her character traits.

8. **Student 2:** Well, her brother's getting more stuff than her. She's feeling left out.

9. **Maggie:** Okay, she feels like things are unfair?

10. **Student 2:** Yeah.

11. **Maggie:** Okay, do you remember the word she uses a lot?

12. **Student 3:** *Injustice.*

13. **Maggie:** Injustice! She feels like she is full of injustice, and nothing is fair to her, and her parents care about everyone else more, and everyone else's parents give more allowance, and she can only buy one hot-fudge sundae a week, and that frustrates her. What else do we know about her?

14. **Student 3:** She's the only girl.

15. **Maggie:** Let's think about character traits. What do we know about her personality?

16. **Student 4:** She's kinda like … she's trying to … she wants to have a better life, living in the museum.

17. **Maggie:** Okay, so if she wants something and sets out to get it, what quality might that show?

18. **Student 5:** Selfish.

19. **Maggie:** She is a little bit selfish, but that's not necessarily a selfish quality. Setting out to get something you want—is that selfish? Putting your mind toward something and really making it work. What might that make you?

20. **Student 6:** Determined.

21. **Maggie:** Determined. So she's a pretty determined character.

22. **Student 7:** She's also thoughtful because she had planned it all out: the trip to the museum and what they can do when they got there.

23. **Maggie:** And *how* they would get there, right? She did a lot of planning. So what might that tell us about her intelligence level?

24. **Student 8:** Pretty high.

25. **Maggie:** Okay, so she's pretty smart.

✔ **Reflection:** How do you help students use evidence from the text to support their feelings and intuitions about characters? How do you help students examine issues from different characters' perspectives?

> ### Maggie's Reflection
>
> *This discussion started off with a great thought, something most kids feel at one point or another—that things are unfair! Since the students could relate to how Claudia was feeling, they had some misconceptions, possibly based on their own experiences. It is important to note that we learned about injustice through Claudia's eyes.*
>
> *Sometimes, a single character's feeling can be unreliable, so we need to confirm using evidence from the text. Students can often use their own experience to make incorrect inferences about situations in books. As a result, this particular topic was a perfect springboard into using evidence directly from the text, rather than overextending students' own connections.*

Identifying Problems

The conversation about Claudia's feelings soon looped back to identifying problems students can solve for the characters. The first problem students choose to discuss is that Jamie only has coins, which he keeps in the pockets of his pants. The coins are problematic for Jamie because they make a lot of noise and weigh down his pants. The class continues to identify problems found in the text, and Maggie lists the problems on an anchor chart (Figure 4.2). Although she is participating in the conversation, Maggie gives equal weight to students' ideas, making sure to add them all to the growing list. Ultimately, the class compiles a list that includes the following problems:

- Jamie and Claudia must remain inconspicuous to adults.
- Jamie only has coins. (The noise from jingling coins may cause adults to notice them. The coins also weigh his pants down.)
- Jamie and Claudia must be able to communicate without words.
- Nobody else can see their things or they will be suspicious.
- They cannot be seen or heard by the museum guards.
- They cannot draw attention to themselves during the day.
- Jamie and Claudia's parents cannot know they have left.
- Jamie and Claudia are short and cannot see over the adults to see the statue.
- There are coins in the fountain that will help them buy things, but they are not sure how to get the coins.
- Jamie and Claudia cannot agree on spending or saving the coins from the fountain.

Introducing the Engineering Design Process

Maggie briefly reviews the class discussion about what engineers do and introduces the EDP. The class discusses individual steps of the process and then decides which parts they will complete during the rest of the period. They decide to tackle Planning, Research, and Design. Maggie reminds the class to think about the problems they've identified so far and prompts them to think about which problems actually need engineering solutions. She has students write those problems on the board. Maggie then tells the class that in addition to

Figure 4.2: Anchor chart of student-identified problems

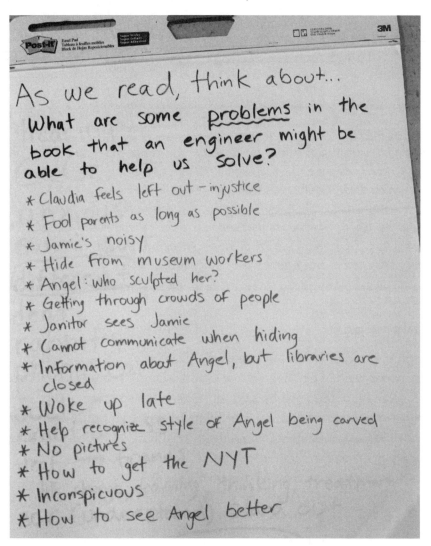

engineering, this project will incorporate reading and writing. Before splitting the class into small working groups, she talks about the expectations of working as part of a team and lists tasks she expects students to complete that day.

> **It is important to present the EDP as a nonlinear process. Think about how you can support students in reflecting on how a design is progressing so they can decide if they need to go back and change their idea, do more research, or use different materials in their design.**

✔ **Reflection:** What expectations would you give your students as they start the task? Are there classroom norms in place that support collaborative work?

Maggie's Reflection

I believe that clear expectations for behavior and tasks when working in groups is important, even when a project is largely student driven. In setting up this project, I knew I wanted to let the students work things out on their own, but to do so, I'd need to help them understand appropriate behavior when working in a group. Therefore, I worked to set clear expectations that were consistent with our previously established classroom routines. I didn't give the students specific roles, but I talked to them about including one another and what it meant to collaborate on a design.

Early Planning

Each working group is asked to discuss the problems on the list and choose one for which they would like to design a solution. Maggie reminds them to stay within the context of the book by saying, "How is the solution going to help the story or the characters, and why might these be important problems to solve in the book?" The groups spend some time considering the different problems. Maggie only gives general guidelines about how to decide on a solution, so students are able to interpret the task themselves.

We will take a closer look at two groups and their design trajectory. The first, composed of Henry and Mitchell, builds a periscope. The second, a group of three girls named Ella, Laura, and Gemma, builds a backpack. Let's begin with Henry and Mitchell as they start to scope the problem.

Henry and Mitchell rank the problems based on solution feasibility, how well a potential solution will meet the characters' needs, and—most important—how "cool" they believe the solution will be. As their discussion unfolds, they begin to find that choosing a problem may involve more than simply choosing a problem from the class list; they must also make assumptions regarding cost, materials, feasibility, and client needs.

1. **Henry:** How about "how to collect more money from the fountain"? [*refers to a problem listed on the board*]

2. **Mitchell:** We could get a bag. [*pretends to put something into a bag*]

3. **Henry:** [*laughs*] How about a padded box? [*pounds fist on table*] A soundproof, padded box.

4. **Mitchell:** [*looks curiously at Henry*]

5. **Henry:** So, we take cardboard—a small cardboard box.

6. **Mitchell:** [*nods*]

7. **Henry:** You know that, like, polystyrene stuff?

8. **Mitchell:** Yeah.

9. **Henry:** We can put that on all sides so it makes it more insulated and harder to hear. And then we could put it on a car ... or on a strap?

10. **Mitchell:** Put it on, um, one of those tiny little ... put it on a miniature car. [*makes hand gestures*]

11. **Henry:** Oh ...

12. **Mitchell:** But we need a water hose and a remote-control car.

13. **Henry:** [*thinks it over*]

14. **Mitchell:** So ... what's the problem? Carrying money from the fountain?

15. **Henry:** I don't know. I don't know if they wanna do ... [*refers back to the characters' objectives*]

In this excerpt, Henry proposes a possible problem for the pair to solve—collecting money from the fountain. He describes an idea for a potential solution as Mitchell listens and evaluates. Mitchell's curious look and interjection that they would need a water hose and remote-control car suggest he is considering the feasibility of the solution given classroom constraints (i.e., it would be difficult to obtain a water hose or remote-control car in the classroom). This cues Henry to step back and consider the clients' needs and desires (line 15), effectively demoting the problem because it does not address the needs of their clients. The two then continue to explore potential engineering problems.

1. **Mitchell:** What about ... I wanna do "communicating." [*selects another problem listed on the board*]

2. **Henry:** Okay, fine.

3. **Mitchell:** We could do, um ... walkie-talkies wouldn't work.

4. **Henry:** [*shakes head*]

5. **Mitchell:** Little headphone walkie-talkies?

6. **Henry:** Yeah, but they would still end up talking.

7. **Mitchell:** No, you could put headphones *inside* a walkie-talkie.

8. **Henry:** Yeah, but then it'd scar my eardrum from talking.

9. **Mitchell:** No ...

10. **Henry:** Yeah, because they would have to talk.

11. **Mitchell:** [*covers mouth with hands and pretends to talk*]

12. **Henry:** [*laughs*]

13. **Mitchell:** Okay.

In this exchange, the pair moves to a new problem on the board and evaluates the potential solution of "headphone walkie-talkies" based on constraints imposed by the book (line 6: "but they would still end up talking [and be heard by the museum guards]") and their personal experiences with headphones (line 8). The boys recognize that this potential solution may not effectively solve the problem for the characters nor be feasible, so they consider one more possibility that involves being able to see a statue, Angel, that is part of a temporary exhibit which is usually blocked by tall adults.

1. **Henry:** Um. How about we try the, um, "How to see Angel"? I was thinking we could, um ... I know how to make it, but I was thinking we could do a periscope.

2. **Mitchell:** Yeah, we could ... [*gestures to mimic holding a periscope*] Yeah, but they [the museum patrons] could still see the periscope.

3. **Henry:** Yeah, but it's not like they wouldn't have to hide ... [*shrugs*]

   [*Both boys appear to imagine air vents and pretend to crawl through the air vents.*]

4. **Mitchell:** Okay, the Angel one sounds cool. That's our final one.

5. **Henry:**   Are you sure? Okay. [*writes while talking*] How to see Angel.

6. **Mitchell:**   How to see Angel.

Mitchell quickly evaluates the idea with criticism since the periscope would be seen by people in the museum (line 2), which he believes may potentially violate the characters' need to remain inconspicuous. Henry then suggests that there is no way to know with certainty that they would be seen; therefore, it may still be plausible (line 3). With little additional discussion, they decide to pursue the problem of "how to see Angel" with a periscope in mind as a possible solution, which Mitchell affirms "sounds cool."

> ✔ **Reflection:** What factors do you think the boys were considering as they picked a problem? Can you point to any moments you think reflect their literacy skills?

In this conversation, Henry and Mitchell explore an open-ended and ill-structured problem, identifying and defining constraints imposed by the book and their classroom. The students consider multiple engineering problems, shifting quickly between imaginative solutions and an evaluation of those potential solutions. For example, Mitchell's argument against the idea of a remote-control money collector introduces the constraints of designing within a classroom and with available materials. In the transition to his walkie-talkie solution, Mitchell relegates the constraints of classroom feasibility, and Henry critiques the design because it could hurt his (or any user's) ears. Lastly, Henry moves to a third problem and suggests a periscope as a potential solution. Mitchell argues that this solution might not meet the objective of maintaining the characters' need to be inconspicuous, yet Henry recognizes this possibility and attempts to come up with an argument in favor of his proposal. At this point, it seems that both boys realize the time constraint inherent to classroom activities and decide to go with the periscope idea. For them, the periscope in an optimal solution that will be feasible in both the classroom and story contexts and be functional under certain conditions of the story.

During their conversation, Henry and Mitchell balance design constraints and criteria while trying to interpret an undefined, messy task. They consider their clients' perspectives and needs as they work to design a functional solution, and they think about how their clients will use the design and how the design could be made in the classroom. Notably, the students listen to each other, build on each other's ideas, and disagree with each other in productive ways. When they push back on an idea, they support their disagreement by referencing the constraints of the classroom and text. Equally important, they are receptive to each other's disagreements, which allows them to move forward and come up with a shared idea for a feasible solution.

Maggie's Reflection

When the boys first mentioned a remote-control car, I worried that they were more concerned with making a "cool" product than making one that would really help the characters. It also seemed like they were overcomplicating the solution in a way that did not address the initial problem, which was collecting money—not transporting it. I was glad (and impressed) that they referred to their knowledge of the characters in order to decide on their problem and solution.

Another observation I found noteworthy was how the boys listened with clear intent to understand each other. They used accountable talk to acknowledge and confirm each other's ideas. As a result, the boys were able to argue and debate in a respectful, productive manner because of their ability to listen thoughtfully to each other throughout the process. They also checked in with each other and reviewed their ideas daily. Ultimately, their ability to communicate, cooperate, and compromise culminated in a solution to an important problem in the book that was achievable within the constraints of the classroom.

Planning

After students have chosen a problem and discussed possible solutions, they plan a design for one of the solutions and discuss the materials they would need to build that design. Maggie provides students with a planning sheet to communicate her expectations and help students structure their thinking. Although students are required to complete the worksheets, the room is full of animated discussions between partners.

As they plan, Mitchell and Henry consider the materials they need for their periscope. They are torn between using cardboard and wood to build the frame. In the following conversation, they evaluate both materials and consider the costs, structural soundness, and feasibility of design for each.

1. **Henry:**　　Do you want to make this out of wood?

2. **Mitchell:**　Hmm ... wood would be more artificial, but it would take longer [to make].

3. **Henry:**　　It would take longer [to make], but it would be stronger.

4. **Mitchell:**	But how would, um, they ... how would *they* get the wood? [*refers to Jamie and Claudia*]
5. **Henry:**	Do they have to?
6. **Mitchell:**	Yeah, but ... you know how Jamie is really cheap?
7. **Henry:**	He is.
8. **Mitchell:**	So, they wouldn't probably get the wood. They would probably get cardboard beause ...
9. **Henry:**	Yeah. I see what you're saying. I see what you're saying.
10. **Mitchell:**	'Cause Jamie's cheap and that would probably cost a lot more than cardboard.
11. **Henry:**	But then cardboard wouldn't be as sturdy, and um, you know how flimsy cardboard is? [*shrugs*]
12. **Mitchell:**	But then they ... once they get the wood, they'd have to get the cardboard. They'd have to get glue. They'd have to get all this other stuff.
13. **Henry:**	Why not just nails?
14. **Mitchell:**	Yeah, but nails cost money too. We could still use wood.
15. **Henry:**	I don't think it has to be something that's in our class.
16. **Mitchell:**	Okay.
17. **Henry:**	I think that's ... I don't think it would be like they have to pay for it.
18. **Mitchell:**	Okay.
19. **Henry:**	[*draws*] So, I think it's just something they would have.
20. **Mitchell:**	Okay.

Throughout this conversation, Henry and Mitchell continually reframe the activity and try to make sense of what they are doing. They consider design constraints and trade-offs (e.g., structure, feasibility, cost) as they design (see Figure 4.3, p. 78). They use what they know about the characters to make inferences about which material they would use (inexpensive materials since Jamie is cheap).

Along with this information from the text, Henry and Mitchell consider constraints of the classroom (available materials and time allotted for construction) and the durability of their design (cardboard versus wood). Henry suggests wood as a building material and cites its durability and strength as important qualities, even though it would take longer to build. Mitchell is skeptical of Henry's sug-

Figure 4.3: Design considerations of Henry and Mitchell's periscope

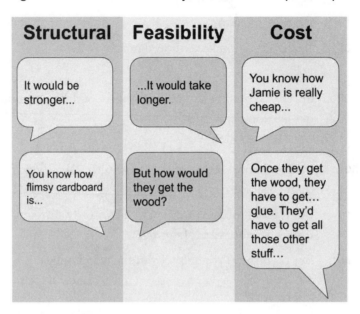

gestion because it would be difficult for the characters to obtain, and he brings up cost, citing Jamie's cheapness, since wood is more expensive that cardboard. Henry counters that cardboard may not be structurally sound.

The partners alternate between referring to their and the characters' needs. As they work, they collaborate and listen to each other's ideas. These are all things that professional engineers think about and do as they design, but instead of drawing from the real world, Henry and Mitchell are drawing from the text. The planning sheet helps them structure and communicate their ideas in multiple ways (see Figure 4.4), thus making planning a multimodal and social activity. The drawings they create as part of this process are also indicative of the visual nature of engineering.

Figure 4.4: Henry and Mitchell's planning sheet

Name _____ Partner(s) _____

The problem we are going to design a solution for is _how to see Angel._

We chose this as our problem because _Jammie and Claudia need to get a better look at Angel to see who sculpted her._

Our plan: _Our plan is to build a periscope-like device to see above the heads of adults and around obstacles. But the mirrors could rotate._

Sketch of our Design

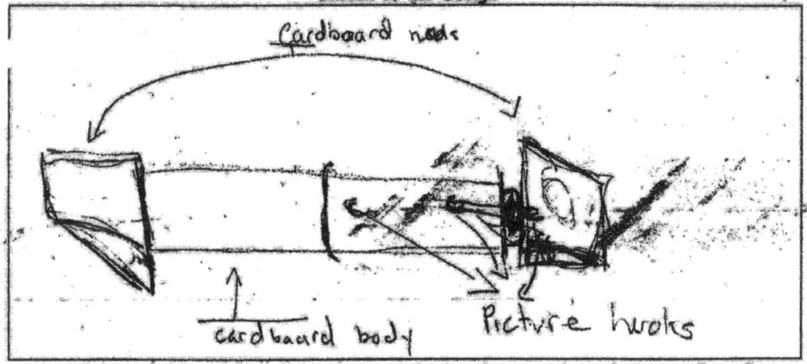

Cardboard nods

cardboard body

Picture hooks

Materials we anticipate needing (Include how much of each material you think you will need):

- 5 medium sized cardboard packing boxes
- 2 6 by 6 cm mirrors
- 6 picture hooks
- 5 cm of stiff plastic tubing
- 2 roll duct tape — red
- 4 thick rubber bands

Maggie's Reflection

Something that stood out to me is how this project made the book more accessible to my students who do not perform well on traditional reading assessments. Mitchell had struggled in other reading assignments in my class, but here he was the one who was referencing Jamie's characteristics and arguing for different materials.

As seen in this exchange, students used information in the text to set the constraints and criteria for their solutions. In addition, the students knew their success depended on how well they met their clients' needs, and their clients were characters from the book. Therefore, in order to be successful, they had to study the characters.

What was notable to me is that students who typically struggled in reading were often the ones who used evidence and inferences grounded in the text to evaluate their design decisions. This gave opportunities for students like Mitchell to feel success in reading because of the Novel Engineering approach!

✔ **Reflection:** How do you see a Novel Engineering project engage struggling readers? How could students' design decisions help you assess their understanding of the text?

While students work, Maggie walks around the classroom and talks to her students. As a "teacher observer," she asks students questions about their design decisions and how their designs would work. Her goal is to better understand their ideas in relation to their engineering and their comprehension of the text, and to help them reason through an problems they may be encountering.

Maggie's Reflection

My role as a facilitator is to ask questions that encourage students to clarify their ideas and troubleshoot through challenges. Furthermore, I model active listening by repeating ideas back to students and asking questions to strengthen understanding and communication. The most important job of the teacher as observer, however, is to listen and learn without judgment. Students cannot surprise or impress us if we deny them the opportunity to do so. Just because a task or project seems implausible to an adult does not mean that it is. Usually, students can and will find a way to make it work; we simply need to allow them to try.

✔ **Reflection:** How is the teacher observer role similar to or different from the role you usually take in the classroom? What are some challenges you might experience in this role?

Building

The questions on the planning sheet allow students to discuss not only their design plans but also the motivations behind their design and building choices. They are meant to help students think in depth about what they are doing and clarify their ideas.

1. **Maggie:**   Can you tell me about your picture?
2. **Henry:**   Um, so, it's like a periscope, but the heads can rotate so you can see in all directions, and it can elevate. We also have a lock, so you can lock it without it, like, swinging everywhere.
3. **Maggie:**   So it doesn't go whacking down on your head or anything.
4. **Henry:**   And we, um, we're gonna try to make it so it can go up and down.

Henry and Mitchell have designed a telescoping periscope that can rotate (see Figure 4.5, p. 82). They are given materials at the beginning of the next class and then spend two sessions building a prototype. They periodically test their design and make tweaks. Upon completion, they present their fully functional design to the class. As they present, the rest of the class asks questions and offers comments and suggestions. Maggie lets the students lead the presentation but coaches Henry and Mitchell through a demonstration of their periscope. Below is a snippet where Henry fields some of Maggie's questions.

1. **Henry:**   Our problem was how to see Angel better, and we built a periscope with some improvements. It has handles, so you can carry it around. You see? And then you can lock it with the mirrors so when you're carrying it, they don't flop. And then to make it less suspicious, you can lock it, so it sort of looks like a lunch box and it's hollow, so you can put food in it.
2. **Maggie:**   So tell us what you see.
3. **Henry:**   [*demonstrates how to use the periscope*] I see Julia. Oh, I see Liam.
4. **Maggie:**   Can you see my hand?

5. **Henry:** I see Carly.

6. **Maggie:** Hold up a certain number of fingers. Put your ... put as many fingers up as you want, Julia. [*Julia stands at the back of the room and holds up a few fingers.*]

7. **Maggie:** So, they have gotten over the fact that they're kids? So now you can see Julia?

8. **Henry:** Yeah, I can see the cursive thing. And now I can see the ceiling.

9. **Maggie:** How many fingers is Julia holding up?

10. **Henry:** Three.

11. **Maggie:** Nice.

Henry shares that they have designed the periscope to look like a lunch box so the characters remain inconspicuous. They then demonstrate how their design works. Henry shows that their periscope allows them to see how many fingers another student is holding up at the back of the room.

Figure 4.5: Henry and Mitchell's periscope design

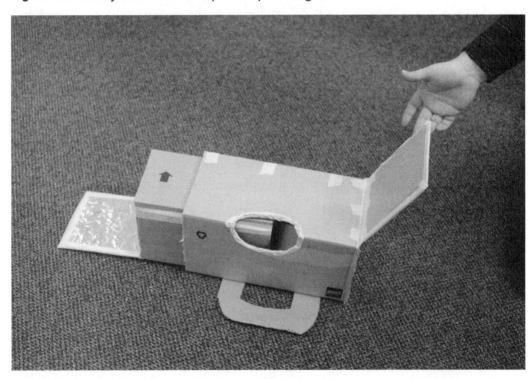

Maggie's Reflection

During this project, I witnessed a ton of communication. Students often have a picture in their head of what they mean but struggle to communicate it to peers. We see this in writing all the time when a student's narrative does not match his or her imagination. In engineering, students need to find a way to make sure their partners understand what they say in order to complete the task. During their discussion, students use spoken words, drawings, hand gestures, and demonstrations. Peers ask questions in order to make sure their understanding matches their partner's ideas. As a teacher, I may need to support students here and there, but when they are invested in their ideas, they will find a way to make sure they communicate it clearly. It's fun to watch!

✔ **Exercise:** The best way to support students doing Novel Engineering is to pay close attention to what they are saying and doing that communicates their engineering and literacy thinking. Recognizing students' productive beginnings can be challenging, so as you read about this group's process, make a note of moments where you see engineering and literacy.

Planning

We are going to move to the second group of students after they have chosen the problem of Jamie and Claudia storing and carrying $25 in change as they hide in the museum. Ella, Laura, and Gemma have decided that one of their design constraints is the fact that Jamie and Claudia must not draw attention to themselves while in the museum. They must remain inconspicuous, which is an idea that runs throughout the book and was a main focus of class discussions.

This constraint leads the students to design and construct a backpack to help Jamie and Claudia hold the collection of coins without being noticed as out-of-the-ordinary museum visitors. As we join the conversation, Maggie is checking in with the group after they have chosen the problem and are trying to figure out what to do first.

1. **Maggie:** What do you think is a good first step that we might be able to get done in one day?

2. **Ella:** Plan.

3. **Maggie:** Plan. All right, great. I think something else also tends to go with planning. When you make a plan, you sometimes have to do what else?

4. **Gemma:** Research.

5. **Maggie:** Okay, good. What else might you have to do? Gemma?

6. **Gemma:** Design.

7. **Maggie:** Design. Do you think we will have time to create, construct, and build right now?

8. **Students:** No.

9. **Maggie:** Why wouldn't we be able to do that right this very second?

10. **Ella:** Um, because planning and researching and designing tend to take a lot of time.

As they plan, the students also discuss design constraints and a feature that will help solve more than the identified problem. Their backpack design will have three compartments: a large one that will be a place to keep clothes and snacks and two smaller ones for change—one that is for spending and one that is for saving. In the following conversation, Maggie is talking with them as they are filling in the planning sheet (see Figure 4.6).

1. **Ella:** So, maybe it goes over Jamie or Claudia's back.

2. **Laura:** [*writes and draws on the planning sheet*]

3. **Ella:** Or maybe it's roundish.

4. **Laura:** And there's a piece of cardboard that separates the compartments so there'd be two—one for saving and one for spending. And on the top, it could hold clothes and snacks.

5. **Maggie:** What problem are you guys looking at?

6. **Ella:** How to get and save money.

7. **Maggie:** What did you come up with?

8. **Laura:** So, we designed this small backpack-size bag that would hold coins because they steal change from the fountain to get money. They don't have much, but they need to hide that money, so they put it in a bag with a compartment.

9. **Ella:** And all the rest of the coins could fit in there.

10. **Laura:** And there are three compartments: one would be for money they could spend, one could be for money to save, and then the one on the top they could use for clothes and snacks.

Figure 4.6: Ella, Laura, and Gemma's planning sheet

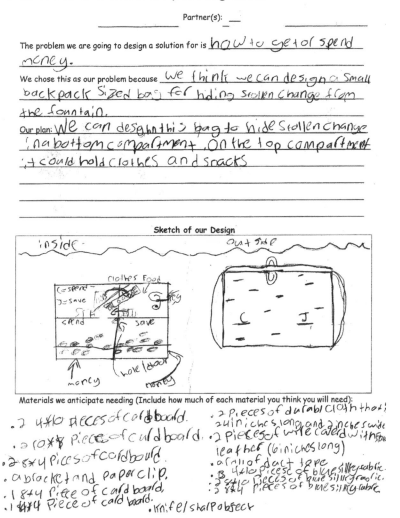

Partner(s): ___

The problem we are going to design a solution for is how to get or spend money.

We chose this as our problem because We think we can design a small backpack sized bag for hiding stolen change from the fountain.

Our plan: We can design this bag to hide stollen change in a bottom compartment. On the top compartment it could hold clothes and snacks

Sketch of our Design

inside   outside

clothes Food

(=spend
)=save

spend   save

money   hole/door   money

Materials we anticipate needing (Include how much of each material you think you will need):

- 2 4#10 pieces of cardboard.
- 2 10#8 piece of cardboard.
- 2 8#4 pieces of cardboard.
- a bracket and paperclip.
- 1 8#4 piece of card board.
- 1 4#4 piece of card board.

- 2 pieces of durabl cloth that 24 inches long and 2 inches wide
- 2 pieces of write cavered with leather (6 inches long)
- a roll of duct tape
- 3 4#10 pieces of blue silky fabric.
- 3 8#10 pieces of blue silky fabric.
- 3 8#4 pieces of blue silky fabric.
- knife/sharp object

The students could have designed the backpack with one compartment for change, which would have addressed the initial problem of keeping the money quiet, but they chose to tackle the additional problem of disagreements between the two characters. Their reasoning is that having two compartments will keep Jamie and Claudia from fighting over how much money they should spend, which is a common conversation between the two characters. The group had identified this earlier as a possible byproduct of their design. They also designed a third, larger compartment for Jamie and Claudia's personal belongings so no one would become suspicious of two children walking around with extra clothes.

Building and Presenting

Ella, Laura, and Gemma test their backpack design with books to make sure it can hold weight. When they are ready, Ella and Laura present their physical prototype to the class (see Figure 4.7). They show the features of their backpack and that it has a false bottom and is insulated to cut down on the sound of the coins. As discussed in Chapter 3, the girls have identified being inconspicuous as one of their design criteria of their backpack. Let's look back at their conversation with their classmates, which also illustrates how peer review can provide another form of information for testing the validity of a design solution.

1. **Ella:** Okay, so our problem was how to hold and save money. Because in the book when Jamie held all his money, it was weighing his pants down and it was making noise, so when they had to go to the fountain, they would only put it in their pockets. And their pockets might hold all the money or they might have to hold it with something else, so we made a backpack that has two compartments. So we are making a small backpack that they can put the money in and nobody will see it.

2. **Student 1:** Won't that still make noise—the change?

3. **Ella:** There's padding.

4. **Student 1:** Would people be able to see from the top?

5. **Laura:** No, because there's going to be a cover here. [*gestures to the top of the backpack*]

6. **Student 3:** I have a question. Why did you cover it with fabric?

7. **Ella:** Oh, because if it was just cardboard, people would notice that cardboard was on their back and they would be suspicious.

8. **Maggie:** What's the word?

9. **Ella:** *Inconspicuous.*

10. **Maggie:** Good, so they want to be inconspicuous.

11. **Student 4:** Who's going to wear that?

12. **Ella:** Both of them. Well, we didn't want to make it a purse because Jamie wouldn't …

13. **Laura:** If Jamie was going to hold it, that would be a little weird, so … [*smiles*]

14. **Ella:** So girls could have worn this, too.

Figure 4.7: Testing functionality of museum backpack during a final presentation

In quick succession, four classmates raise questions regarding the functionality of the backpack design. When Kate suggests that the coins will still make noise in the backpack, Ella points right away to the padding designed to absorb that sound. Eric then raises the concern that people would be able to look into the backpack and see its contents, but Laura jumps in to describe a design revision they have planned but not yet implemented: a concealing lid. Eric's question confirms that their revision plan is a valid one. Carola then asks about the fabric cover on the pack, and Ella explains that this design choice is a way to meet the constraint of inconspicuousness. Finally, Gabbie's question about which of the two characters would wear the backpack gives the group a chance to test out their idea that the backpack is appropriate for either a boy or a girl. The classmates' willingness to raise questions about the backpack design indicates an awareness that input can be used to affirm or alter the course of design.

Maggie's Reflection

I think the feedback process was incredibly important both to the engineering design process and to students' feeling of success because the criticism and questions came from their peers. These types of projects rely on the communication of ideas. Students work so hard at communicating ideas to their partners that the groups end up with a shared vision of their creation. They have to repeat this process as a team when presenting to the group, and they need to be willing to hear criticism as a means to improve.

During this process, they also need to convince their peers that their design is plausible, defend specific choices they've made and how those choices affect the prototype, listen to suggestions, and recognize how those suggestions could improve their design. When suggestions come from their peers and friends (rather than from a teacher or adult), they become much more meaningful and a cause for perseverance. Often, when adults are too present, students fear being wrong, so they stifle their ideas to avoid that possibility. That fear seems to disappear when students face their peers.

Many of the engineering problems students noted were somehow related to Claudia and Jamie's need to be inconspicuous. Initially, students started discussing the characters not being seen or making sure they weren't found. As a teacher, my job was to incorporate vocabulary from the text. Each time the topic came up, I made sure to use the word inconspicuous *and encouraged students to use it as well. Ultimately, we used the word so often when discussing the book that it became part of the class vocabulary.*

✔ **Reflection:** If you model how to provide critique and ask questions, you help students understand the expectations for presenting and being engaged members of the audience. As you read through this example, what questions could you imagine having for the group?

Writing and Assessment

Throughout this unit, Maggie is able to assess her students' understanding of the text and the engineering. She makes sure students are able to present their knowledge in a variety of ways, including written assignments, oral presentations, and small-group and individual conversations.

As a final component to this Novel Engineering unit, Maggie asks her students to write an alternative ending to the book—one that includes their design and how it would affect the story. The students' new endings give Maggie additional insight into students' understanding of the text. In the following exchange, Maggie helps Gemma frame the writing activity.

1. **Gemma:** So, Jamie has a lot of money in his pockets, and it's weighing his pants down, but then they remember they have a backpack and that's our invention: the backpack. And so, then they start using it. And then ...

2. **Maggie:** And then it goes from there?

3. **Gemma:** Yeah.

4. **Maggie:** Is there going to be suspense?

5. **Gemma:** I don't think so. Like, we could add suspense.

There is considerable chatter in the room as students talk to their partners and plan what to write in their alternative book endings. (See the Extras page for this book at *www.nsta.org/novelengineering* for examples of student work for this assignment.) Reading the new endings, Maggie is able to assess students' understanding of the text and how their design should function while working on writing objectives she has outlined for the class.

Maggie's Reflection

Once again, this writing helped focus the engineering process around the book. In order to write an alternative ending, the students needed to understand how their design would affect the characters and how that effect would be a catalyst that might alter the story itself.

The initial challenge proposed to students was to invent something that would solve a problem the characters in the book faced. This writing assignment required them to evaluate their invention in terms of how it would fit into the book and how the characters would use it. The evaluation and writing processes helped me assess my students' thinking because their stories were essentially reflections on the entire Novel Engineering process. It was a perfect culmination to integrating literacy skills with an engineering project.

Having said that, I was pleasantly surprised by the inventions my students created. I was also particularly impressed with how much the students discussed the characters throughout the process. Sometimes, we overemphasize using the actual text, but it was clear that the students knew the two characters as well as they might know their friends without having to go back into the text and find evidence. They were able to cite evidence from memory because they were so connected with the project.

Wrap-Up

In these case studies, we showed what it looks like for students to engage in their first Novel Engineering experience and provided the teacher's reflection at different times throughout the activity. What surprised Maggie (and us!) was the fluidity of students' engagement in engineering. They easily slipped in and out of the story, using their knowledge of the characters and the plot to inform engineering decisions, and they navigated the engineering design process as they developed their solutions. We were also amazed by the complexity of their thinking. Students carefully considered aspects of the story and their classroom, and they even made evaluated trade-offs when making design decisions.

Safety Notes

1. Wear safety goggles or glasses with side shields during the setup, hands-on, and takedown segments of the activity.

2. Use caution when using hand tools that can cut or puncture skin.

3. Use only GFI-protected circuits when using electrical equipment, and keep away from water sources to prevent shock.

4. Secure loose clothing, remove loose jewelry, wear closed-toe shoes, and tie back long hair.

5. Wash your hands with soap and water immediately after completing this activity.

Reference

Konigsburg, E. L. 1967. *From the mixed-up files of Mrs. Basil E. Frankweiler*. New York: Atheneum Books.

Book Resource

From the Mixed-Up Files of Mrs. Basil E. Frankweiler; Konigsburg, E. L.; Age Range: 8–12; Lexile Level: 700L

Helping Trudy Swim the English Channel

Chapter 5

> As you read this chapter, think about the balance between students' ownership of a project and pushing them to create a functional and useable solution.

Novel Engineering differs from most school design projects because it places significant responsibility on students to scope a problem based on their reading of a text. For many Novel Engineering projects, students identify the problem, determine the constraints, and decide criteria for success. In other words, they do much of the work of figuring out what counts as a solution. This freedom offers opportunities for students to be creative in how they think of problems and solutions.

At the same time, however, we want students to balance their creativity by considering how their solutions will work—both in the classroom and in the context of the story. Teaching engineering design, therefore, involves balancing students' agency and ownership of the project while also holding them accountable to functionality and usability. Furthermore, these considerations are on top of many others when teaching elementary school students (e.g., helping them learn to communicate, manage their emotions, and feel confident in their abilities).

The case study in this chapter explores how this balance plays out in one third-grade classroom. We look at both how students take up the task of designing solutions and how the teacher responds to help shape their understanding of the task. In this classroom, students are solving problems faced by the main character in the nonfiction text *America's Champion Swimmer: Gertrude Ederle* by David A. Adler. After the class reads the text and identifies problems, the teacher, Anne, encourages her students to brainstorm many creative ideas for solutions. Then, as the students narrow down their solutions and begin building, Anne helps them shape their engineering so students can both describe how their designs meet their client's needs and demonstrate how their prototypes work. Alongside these negotiations, Anne manages other challenges students have when communicating and collaborating with one another.

To chronicle the class's trajectory, we describe different patterns in how students interpret the task and how their interpretations change over time, particularly in terms of what it means to design and build a solution. To this end, we zoom in on one group that made remarkable progress; students in this group began with "crazy ideas" but ended up with a functional solution they could demonstrate to the class. They could also describe what modifications were needed for the main character, Gertrude (called Trudy), to use the design. Along with these descriptions of students' activities, Anne offers her own perspective, describing what she notices about her students' engineering and why she responds as she does.

Anne's Reflection

At the time of this case study, I was a ninth-year teacher, mostly with third and fourth graders. I was excited to do Novel Engineering in my classroom because it was a way to teach reading comprehension in a new way as well as introduce engineering to elementary students. In particular, I hoped these projects would help students who didn't always feel successful in school by giving them opportunities to show their strengths. For other students, I thought these projects would present new challenges. With engineering, I especially liked the idea that students could learn to fail, meaning they would learn that it's okay if you fail, and that you can grow from the experience. At the same time, I appreciated that Novel Engineering projects also support students' literacy, including their reading comprehension and communication skills.

✔ **Reflection:** Anne's reflection that Novel Engineering has multiple possible solutions means that as an educator, she has to recognize and support many different pathways and solutions. Are there experiences you've had as an educator that help you think about supporting teaching with multiple solutions? What are some challenges you might anticipate?

Background

This case study takes place in a third-grade classroom in a rural school district in Massachusetts. When the school first agreed to participate in Novel Engineering, the district also adopted Reading Street, a standardized literacy program based on anthologies. At the time, most Novel Engineering projects used children's novels, so we were curious to see how integrated engineering and literacy units would work with shorter texts (such as those found in the Reading Street curriculum).

Two third-grade teachers (including Anne) identified *America's Champion Swimmer* as a promising text within Reading Street for Novel Engineering. This short book tells the story of Trudy, an Olympic swimmer who set records for her long-distance swims. The biography chronicles the challenges she faces in becoming the first woman to swim across the English Channel.

The teachers were excited that the engineering client would be a real person who faced real problems—as opposed to fictional characters who responded to fantastical elements. The case study we explore here is from Anne's second implementation of Novel Engineering using this text. It was the first engineering project she did with this class, though, and it was a new experience for many of her students.

> ✔ **Reflection:** What are the trade-offs between reading full books versus portions of a text? How much do students know about the setting and characters? How does this influence their ability to design for the characters?

Anne's Reflection

As a teacher, I expected to facilitate Novel Engineering projects much like other classroom activities in which I helped students arrive at a particular understanding—for example, a particular mathematical procedure or definition of a vocabulary word. However, in my first time teaching a Novel Engineering project, I realized there are lots of ways of thinking about a problem and multiple possible solutions for each one. Furthermore, I realized that part of what I wanted my students to learn was how to try out their own ideas, even if they don't work.

(continued)

(continued)

For example, I remember talking to one group about their design—I knew their idea would't work, but they didn't realize it yet. I kept questioning, but that both confused and frustrated them. My role as a teacher had to shift. Since then, I have tried to step back more and let students try out their ideas, even if they fail. I'm learning to balance letting them have ownership of their ideas while also holding them accountable for being able to show how their design would work, both in the classroom and in the story.

✔ **Reflection:** What strategies do you already use with your students that could also be used to support collaborations and encourage students to listen to one another?

Timeline of Project

The students read *America's Champion Swimmer* twice over the course of two days. The engineering portion of the unit takes place over three days, with students spending approximately two to three hours each day identifying problems, planning and realizing design ideas, and then iterating and testing their solutions. We present a timeline of these activities in Figure 5.1.

Identifying Problems and Brainstorming Solutions

The class begins the project by reading *America's Champion Swimmer* aloud, and students listen for problems Trudy faced. Anne stops at various points to ask about the characters and plot, such as what it means to be someone who accomplishes a challenge for the first time.

As the class rereads the text, partners work together to read for and brainstorm a list of problems Trudy faced. Once the class generates a list of problems, Anne asks them which problems could be solved using engineering. This class focuses on the engineering problems that arose when Trudy swam the English Channel, and they identify three for which they think they can design solutions:

- Helping Trudy eat and drink
- Preventing water from stinging Trudy's face
- Helping Trudy see her guide boat

Figure 5.1: Sample Novel Engineering classroom timeline

Day 1	Day 2	Day 3	Day 4
Reading the text aloud	Reading the text in pairs	Sharing in small groups	Testing and redesigning prototypes
Discussing characters and plot	Identifying engineering problems	Planning and building prototypes	Sharing designs with the class
	Brainstorming solutions individually	Getting feedback from peers	

Anne then asks students to work independently and brainstorm possible solutions for each problem. Students also rank the problems they want to work on. Finally, Anne organizes students into groups of three or four by partnering students who work well together, and she assigns each group one of the three problems (based in part on their rankings).

Conceptualizing and Realizing Design Ideas

Before organizing students into groups, Anne talks about the different ways they could collaborate to come up with their designs. For example, they could use just one person's idea, combine several ideas, or come up with a new idea. She also reminds students that they can change their designs later. One suggestion Anne gives so that all student ideas get heard is for each group member to share his or her ideas one at a time. Once everyone has shared, groups can then discuss which idea they want to pursue.

Anne's Reflection

In this particular class, there was a lot of diversity in how outspoken the students were and how well they could work on teams. Therefore, I spent a lot of time planning how to support their collaborations so they were listening and responding to one another to come up with a solution.

Anne then reviews a planning worksheet for the group (see Appendix I, p. 236), detailing the different sections that ask students to describe their idea, sketch their design, describe how it works, and list the different materials they need. Anne places the materials (e.g., plastic bottles, cardboard boxes, aluminum foil, paper cups, tubing) on a table for students to look at, but groups must decide on a plan before they start gathering supplies. Anne emphasizes that she wants students to focus on the process of designing, working together, and being flexible with their design idea.

> ✔ **Reflection:** Think of your own students. How could you pair students with various learning styles and skill sets so all students are able to access the activity and work together?

Coming Up With Ideas

To see how students engage with the project, we will look closely at the design process of one group, composed of students James, Sophia, Amber, and Ella. Anne grouped these students together because two of them are typically more vocal and two are often very quiet. She wanted the group to represent diverse student interests and abilities. When they first meet as a group, James jumps right in to share his brainstorming.

1. **James:** Okay, okay. Look! I had so many crazy ideas! So I was thinking of a helicopter up in the air, dropping a table on ropes that had, like, food on it. Okay, and one of my other ones is a feeding tube, out of a boat. Another one is like a table coming out of a boat.

2. **Sophia, Amber, Ella:** [*giggle*]

3. **James:** Okay, and another one is like a hat that's like a table on top …

James starts by sharing lots of ideas. Based on his possible solutions, he is thinking broadly about the problem. His ideas incorporate mechanisms for delivering food to Trudy and for how she would eat it. It appears that the group recognizes these ideas are not entirely feasible; James himself calls his ideas "crazy," and the other group members laugh at his suggestions.

Anne's Reflection

I really appreciated how James was having fun and being goofy but still being thoughtful and coming up with lots of possible solutions. The class and I had watched a video of NASA engineers discuss how they had started with "crazy" ideas, such as sending a monkey to space—and that some of those ideas were realized!

When we started this project, I told students to think of lots of different things first, and maybe a piece of one idea would work. It didn't surprise me that James had so many enthusiastic ideas, but I was also glad to see that he later made sure everyone got a chance to share. I had deliberately asked students to brainstorm solutions independently and write down their ideas.

When they broke off into groups, I told them to listen to everyone's ideas first. My goal was for students to work together to decide which idea to pursue, plan a solution, and determine its feasibility. I set up this structure so that the quieter students would have space to contribute at the start. I think this group did a great job listening to and respecting one another's ideas even at the very start of their collaboration.

✔ **Exercise:** As you read this excerpt, pay attention to students' ideas about testing and the mechanistic ideas about how the solution will work. See if you feel that you understand their ideas and if their tests will show them if their designs will work.

Planning

After all group members have an opportunity to share their ideas, they decide to pursue James's tabletop hat idea. They then turn to the planning worksheet.

1. **Amber:**   Okay, so we'll go over the materials, figure out the materials. What do we do now?

2. **James:**   We have to write the possible solution.

3. **Amber:**   Possible solution.

4. **James:**   Sophia, you do it.

5. **Amber:**   Wait, how are we gonna do the solution? Oh, yeah, we—

6. **James:**   We gotta do that hat …

7. **Amber:**   We get a hat—

8.	**James:**	That you put on! Just, like, describe it. It's a hat with a table on the top.
9.	**Amber:**	Well, I was thinking of a hat, like— [*starts to draw*]
10.	**James:**	No, I was thinking like a round thing that holds things. [*gestures*]
11.	**Amber:**	How are they gonna test it, though? Pull it off?
12.	**James:**	What do you mean?
13.	**Amber:**	Trying to get the food off the head.
14.	**James:**	No, you just … nosh, nosh, nosh. [*pretends to eat*]
15.	**Amber:**	So, like a hat with food in it? Wait—
16.	**James:**	Yeah, how about with food in it? Like, it could be like a cylinder, and then we would put food on her head.
17.	**Sophia:**	I was thinking like a cardboard hat. We could poke holes in it, with a hole puncher thingy—
18.	**James:**	Hole puncher.
19.	**Sophia:**	Yeah, and get string and tie it in a knot and have bottle caps hanging.
20.	**James:**	And that would be the food?
21.	**Sophia:**	Yeah.
22.	**James:**	Or we could do, like, a cylinder hat. Okay …

The questions on the worksheet prompt the group to flesh out their idea even more. James describes and gestures that the hat is shaped like a cylinder, and Amber suggests that the hat could hold the food inside it. Based on their gestures, it seems they are thinking of the hat as something worn by someone standing upright rather than someone swimming (and therefore horizontal).

Amber makes an interesting comment when she asks how "they" are going to test it (line 11). It's not exactly clear what she means here. One interpretation is that she is thinking of the classroom requirement of "testing" their ideas; therefore, "they" could mean other classmates. Another interpretation is that she is wondering how Trudy would eat off the hat in the water, raising doubt that a swimmer could pull the food from it. Both types of interpretations are often present during student discussions in Novel Engineering, which is why it is important for teachers to speak with their students so they can understand what they are thinking. James responds by pretending to grab food from his head and eat it (line 14).

Sophia then jumps in to suggest that they could make the hat from cardboard and have bottle caps to represent food that hangs from holes in the hat (lines 17 and 19). At this point, the group is starting to get into details of their design, including how they could make it. Sophia describes specific materials they would need and steps they could take to construct their design.

> ✔ **Reflection:** This is our interpretation of the students' exchange. Do you agree or do you have a different interpretation? What else do you think Amber means when she asks how "they" are going to test it? What do you think Amber and Sophia are proposing?

Anne's Reflection

What first stood out to me about this conversation was Amber's question about how they would test their design. It's one of the questions I would've asked if I had come up to them at this point. It doesn't seem like they were thinking of how Trudy would use their device, so I probably would have prompted them to think about swimming and how your body is oriented in the water. Amber's question wasn't as specific, but I think it's great she was critically examining their solution. I was particularly encouraged because she isn't a very confident student, but she felt comfortable enough to question the group's plan in productive ways.

More generally, I was pleased this group was collaborating so well. They were listening to one another and building off their ideas. Though they chose to pursue one of James's ideas, both Sophia and Amber began offering suggestions for what to do next and how to modify the idea. I also noticed that James didn't just listen to Sophia but checked to make sure he understood her idea by repeating parts of what she said.

Noticing and Responding to Other Groups

Anne doesn't spend much time with this group, in part because they collaborate well from the start. This is not so with other groups, however. A few students have a hard time listening to their peers, dealing with conflicts, and coming up with compromises. Anne spends more time with these groups, helping them communicate with one another and deal with their emotions as they work together. This translates into different expectations for the different groups' final products. Instead of pushing all students—particularly those having difficulty working together—to design a functional prototype, Anne's goals are for them to start having productive discussions and solve problems as a team.

Some groups need help figuring out the project more generally. Some students have a hard time imagining a solution that isn't already written about in the book (e.g., swim goggles). Some groups start building what looks like a backdrop for a play, including water, a boat, and a small swimmer. Some groups focus less on their design and more on the materials they want to use.

These different challenges mean that Anne has to tailor her responses (and expectations) for different groups. However, some issues cut across groups, particularly with respect to functionality. Anne ends up telling the whole class that they are not doing "set design" but making something that shows how their idea works—even if it is just a portion of their overall design.

Mid-Design Share-Out

The focus group eventually realizes that their hat idea is problematic for swimming. James recounts their discussion later: "If she's doing a backstroke, then everything would fall off." They instead choose to develop an idea for something Trudy can wear on her stomach. (See Figure 5.2 for a drawing of the group's design).

They present their prototype to the class at the end of the second day of the project—after about one and a half hours of planning and building.

1. **Anne:**   James's group. Your problem was the food?

2. **Students:**   Yes.

3. **James:**   This is like—

4. **Amber:**   It has food in it.

5. **James:**   A container. Part of it is food, and part of it is water, so this would be the one for the food, and this would be for water. So if she got hungry or thirsty, she would just suck out of it. And then the food or the water would just come out of the tube.

The students develop a pack to be worn on the front of the body when swimming. The pack contains both food and water, and Trudy could get both from the tubes. As James describes their design, he is explicit about how their design addresses Trudy's problems. He uses their prototype to describe how it would work. The group has incorporated tubes to represent a more realistic delivery mechanism than strings and bottle caps (their original idea), but their design is still primarily representational.

One classmate in another group asks how they came up with this idea. James talks about how they changed ideas from the tabletop hat to their current design. He then elaborates on how their design could be used by Trudy when swimming.

Figure 5.2: Trudy's back or bellypack

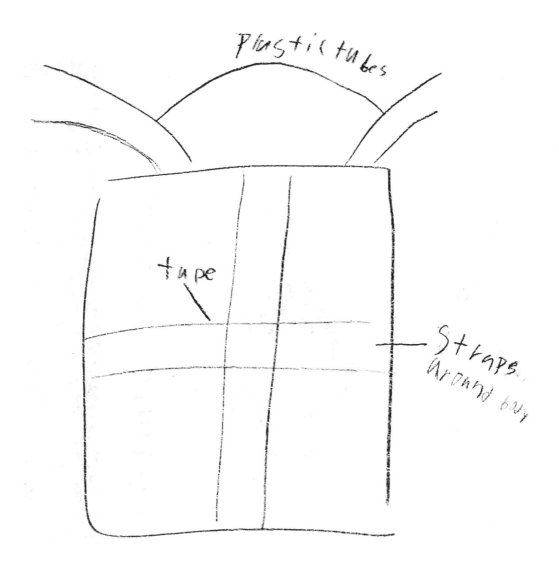

He explains that Trudy could do the backstroke and wear their design on her stomach, and if she were swimming freestyle, she could wear it on her back.

Another student asks how food would fit through the tubes if there were "big pieces of chicken." Although this question may seem silly, chicken is one of the foods that Trudy actually ate during her swim (as described in the text). James responds that the pieces would be chopped small enough to fit through the tubes.

> ✔ **Reflection:** One of the goals of Novel Engineering is to push students' mechanistic thinking and push them to design functional solutions. What questions might you have about the functionality of this group's design?

> **Mid-design share-outs help students refine their ideas and get feedback from peers. What types of questions could you model for your students to help them ask more productive questions?**

Anne's Reflection

Overall, I felt as though the students had a good start on their designs. Everyone had something they could share with the class, even though some of their questions weren't particularly useful, such as asking how they got that idea. I was excited to see students refer to the text to talk about what Trudy ate during her swim. However, there wasn't a lot of focus, either in their presentations or in their questions, about the functionality of their designs.

I remember being concerned that some groups, such as the focus group, thought they were finished and didn't need to improve their designs. I wanted them to realize there were ways to make their solutions even better.

Revising Solutions

Anne starts the next day by writing questions on the board:

1. Can you test your design?
2. How can you make it work?
3. What can you do to improve your device?

In particular, she presses students to think about how they could make their designs better. Some groups want to start over and ask if they could make a large-scale design after making a smaller model. Anne encourages students to improve on their earlier ideas, even if they start work on a new prototype, and show how at least a piece of their design would work.

> ✔ **Exercise:** As you read the following excerpt, pay attention to the students' ideas about how they could change their design.

Revising

The focus group responds immediately to Anne's comments about needing to make their device work. They start brainstorming what to do, but they also seem nervous about taking apart their device to make changes. Sophia suggests they open up the box to get to the tubes, but James protests that it might take too much time to take off all the duct tape. Amber goes back and forth between wanting to start over with a different idea entirely and making suggestions for their current design.

1. **Amber:**   I was thinking we could get a cup and cut the bottom of it and then fill it up with water.

2. **Sophia:**   And put one of the plastic things in that were up there yesterday.

3. **Amber:**   I think we should cut cups at the bottom, besides the one that's going to be filling everything up, and then we could just glue it down and tape it, and make sure it doesn't come off.

4. **James:**   Guys, we need to be—

5. **Sophia:**   I think we need a—

6. **Amber:**   I think we need to start over. We need help. We don't know if we should start over or—

7. **Sophia:**   I think we should just take the whole box open—

8. **Amber:**   And then we could—

9. **Sophia:**   And then cut this. But then it wouldn't work because it's two—it's one tube, going like that. And it needs to be two tubes, going like that.

10. **Amber:**   Okay, so let's do that. Let's take everything off.

Amber and Sophia begin to redesign their prototype since Amber's initial idea isn't very clear at first. She talks about cutting the bottom of cups and filling them with water, but then it seems she is thinking of stacking cups as a way make a tall container for the water. This explains why she wants to cut the cup bottoms, except for the one that would be used as the base to fill everything up.

Sophia suggests cutting the one long tube on their current prototype. As she explains why they need to cut the tube, she gestures that, at this point, the two ends are just connected to each other. With two tubes, she shows that they would be parallel. Her suggestion spurs the group to action and they start taking apart the box. The group spends the rest of the time working on how to make their device functional so that water can be sucked out of the tubes. They start building Amber's cup tower (but ultimately find a larger water bottle to use instead).

Anne's Reflection

My introduction at the start of the second day of building was in part directed to this group. I remember thinking that they considered their design finished, and I wanted to push them to come up with a more functional product. However, I was still surprised (and pleased) at how many changes they made.

I think this snippet shows that their collaboration was critical in helping to support one another as they dealt with the challenges of redesign. Amber and Sophia recognized different problems they needed to solve to make the design functional. Amber focused on holding the water while Sophia focused on the delivery mechanism. Despite attending to different aspects of the design, they listened to each other. Sophia then made their task more manageable by breaking down what they needed to do step by step.

I again was impressed with Amber, who, despite feeling anxious about modifying their device, led her group in coming up with a major change to the design. She was persistent in expressing her ideas, at one point even drawing to show how she proposed to stack the cups. Even though her idea was eventually abandoned, she got to communicate and try her ideas, which was a major accomplishment for her in this project.

Noticing and Responding to Other Groups

As other groups gain a better sense of what it means to build and test an engineering prototype, Anne starts to deal with other issues that emerge as students grapple with making functional designs. Like Amber, some students react negatively to setbacks, and Anne responds to help them deal with these emotions so they can continue working. For example, in one group, a student ends up taking a break from the group work to talk with a paraprofessional. He is frustrated because he feels as if none of his ideas are being used, and he doesn't feel like he can compromise.

> **Some students become focused on the final solutions and feel as if they have failed if their solution does not work perfectly. Are there instances in your classroom when you ask students to focus on the process rather than the product? How could you set up this expectation for Novel Engineering units?**

Other groups cannot make their designs functional—because of a lack of time, because of the materials available to them, or because their ideas are just not feasible. Part of engineering is taking risks and failing, and Anne doesn't want to punish students if they cannot make their designs functional. She tells these students to talk about the design process and their ideas about how their mechanism *should* work. However, she does not allow groups to simply black box their mechanisms. Students cannot just say that Trudy will press a button and food will come out. Instead, Anne presses them to describe, in step-by-step detail, how their devices will function.

Final Share-Out

At the end of project the students present their designs. The focus group has a working prototype to demonstrate to the class. James shows how Trudy would wear it.

1. **Sophia:** She would have this for water, and the tube would come out and she would suck out of it.

2. **James:** But in the real thing, there would be two compartments, one for food and one for water. And then, this is how it would work. [*sucks water up one of the tubes*]

3. **Sophia:** Lift it up so they can see. But this would be closed. [*points to the open box showing the water bottle and tube*]

4. **James:** Yeah, and it wouldn't be a cardboard box because that would get wet easily. It would be a plastic container.

The group shows off a working prototype, and James demonstrates how to suck water out of the tubes while wearing the device on his stomach. However, they also refer to aspects they would change for the device to be used in the book's context. For example, there would be two compartments, and the pack would be sealed and made from a waterproof material like plastic.

When it is time for questions, one of their classmates asked how they got their device to work. Sophia describes the process they followed to revise their design. For example, they cut a hole in their pack and placed a water bottle inside, underneath this hole. They also put a tube in the water bottle and made sure it was long enough to reach James's mouth.

> **Anne contrasts designing in Novel Engineering to other projects she does in the classroom. Are there any projects you can reference to help your students think about designing? What are they and what are the characteristics that will help students frame their thinking about creating functional solutions?**

Anne says that one of the major challenges she's faced as a teacher is not getting too involved in students' projects, rather letting students work their ideas out for themselves. However, as we can see from this case study, that doesn't mean she never intervenes. Specifically, she noticed that many groups were not thinking about functionality and reminded them throughout the project that they need to show how their ideas would work. She also noticed that students were having difficulty figuring out what it means to design a solution, so she contrasted the task with other projects they had done, such as making a diorama or backdrop for a set. Since this case study, Anne has taught several more Novel Engineering units and has found that these issues come up every year, especially for the class's first engineering project.

When talking to groups individually, Anne persistently presses them about their design, especially if they haven't thought about how they will show its functionality. If students have an idea for a mechanism, she generally lets them try out their design. This keeps students from relying on "magic" or creating only representational designs, and it gives them freedom to explore their own ideas. Often, students end up refining their ideas after testing functionality or receiving feedback from peers.

It's also important to recognize that Anne formed different expectations for different students. For some groups, such as the focus group, she found that they started off well as a team, coming up with creative ideas, refining those ideas, and starting to build their designs. She could push them to think about how they could show functionality and consider constraints on the problem. However, other groups had trouble conceptualizing the task and didn't know how to work as a team or think of a solution that wasn't already mentioned in the book. Anne therefore had to tailor her responses to support their engineering. As a result, different students learned different aspects of what it means to be an engineer.

Although communication and teamwork are a focus for students' overall development, this case study shows that these skills are critical for groups to engineer a functional, usable device. Students need to listen to and understand one another's ideas—not just "get along." The students in the focus group questioned one another, made decisions based on the quality of their ideas, and helped one another deal with setbacks. These actions helped them work as a team and ultimately improve the design of their solution.

> ✔ **Reflection:** What are your expectations for your students? Where do you think they will excel and where do you think they will need extra support? What types of supports could you put in place during a Novel Engineering unit?

Wrap-Up

Anne taught the same students the following year and did another Novel Engineering project with them. The students, now in fourth grade, designed solutions to problems identified in *From the Mixed-Up Files of Mrs. Basil E. Frankweiler* by E. L. Konigsburg. By then, Anne knew the students quite well. She had seen how they worked together, came up with ideas, and dealt with failure during both engineering projects and other school activities. Likewise, the students were more familiar with Anne and one another. They also had more of a sense about what it means to design a solution to an engineering problem. This familiarity shaped how Anne planned and organized the project differently from her first Novel Engineering project.

During the fourth-grade project, Anne noticed that the students quickly recognized that their task was to make a functional prototype and solve a problem from the book. In contrast to the first project, she didn't see students making dioramas of the setting of the book or focusing on decorations. Instead, she saw them draw on the classroom's resources and test how their devices might work in the context of the story.

Anne therefore had more time and freedom to focus on identifying the engineering design process and pointing out how students were doing various steps. She reinforced students' work as engineering, pointing out when they were engaged in planning, designing, testing, and iterating their designs as engineers would. This approach stands in contrast to the design of many engineering curricula in which students are told the steps of the design process and then asked to engage in them. Instead, Anne helped her students engage in engineering and then linked it back to the formal representations of the process.

Safety Notes

1. Wear safety goggles or glasses with side shields during the setup, hands-on, and takedown segments of the activity.
2. Use caution when using hand tools that can cut or puncture skin.
3. Use only GFI-protected circuits when using electrical equipment, and keep away from water sources to prevent shock.
4. Secure loose clothing, remove loose jewelry, wear closed-toe shoes, and tie back long hair.
5. Wash your hands with soap and water immediately after completing this activity.

Book Resources

America's Champion Swimmer: Gertrude Ederle; Adler, D. A.; Age Range: 4–7; Lexile Level: 800L

From the Mixed-Up Files of Mrs. Basil E. Frankweiler; Konigsburg, E. L.; Age Range: 8–12; Lexile Level: 700L

Rethinking the Colonial Period

The students in this case study are dealing with how the context of a historical time period influences their designs.

We previously discussed how Novel Engineering activities offer students an opportunity to be creative in how they think about a problem and solution. We also discussed how the open-endedness of Novel Engineering activities requires students to maintain a delicate balance. On one hand, they have freedom to be creative as they scope the problem and brainstorm imaginative solutions; on the other, they must consider objectives and constraints, such as how their designs will work both in the classroom and in the context of the story. For teachers, managing this balance involves sustaining students' ownership of their projects while also holding them accountable to engineering criteria, such as functionality and usability.

In this chapter, we show some of the complexities of maintaining this balance, specifically for students using a nonfiction historical text. In Novel Engineering units, the stories students read may transport them to different times and places, spanning historical eras, geographic locations, and fantastical worlds where animals act like people and characters have magical abilities. As students develop their designs, they are also responsible for showing their teacher and classmates how those designs work. In other words, they have dual objectives:

- To show how their design would work within the story to help the characters;
- To demonstrate how their design functions in the real-world setting of their classroom.

Through our research, we have noticed that these objectives are not always compatible. Students' ideas about how a design might work in a story may include features that are not physically possible to create in their classroom environment, especially if they require advanced technologies or materials that are not easily attainable.

In Chapter 5, we saw how students managed their dual objectives as they worked on a design solution to help Trudy Ederle, the first woman to swim across the English Channel, overcome challenges she faced while swimming. The students recognized that they were designing solutions to help Trudy but were also responsible for constructing a design that would be functional in their classroom environment. At the beginning of the activity, students started out with "crazy ideas," which included designs they would not be able to construct and test in the classroom (e.g., a helicopter dangling a table on ropes, a table coming out of a boat). As they continued to discuss their design ideas, they recognized that some of their ideas might be impossible for Trudy, and they would certainly be impossible to build in the classroom. The students continued to frame the problem and narrow the possible solutions until they converged on the idea of a special backpack that would allow Trudy to eat and drink while she swam.

The students managed their dual objectives by showing a functional prototype to classmates and describing the features that would be different for Trudy. To show functionality, they demonstrated to their classmates how their design worked by sucking water through one of the backpack tubes. When they presented their design, they described how it would help Trudy. For example, they said, "She *would* have this" and "The tube *would* come out and she *would* suck on it." They then added that it would be different "in real life" and that "it wouldn't be a cardboard box because that would get wet easily."

A Design for Colonial Times

In this chapter's case study, we take a look at third-grade students engaged in a Novel Engineering activity based on a nonfiction historical text. We show how students progressed through the design process and occasionally struggled and argued—specifically about materials that would be functional in the story and classroom settings. We describe how their arguments, though tense at times, served to help students better understand their own design objectives.

> Learning about colonial times via Novel Engineering provided
> students with a greater understanding of the era than if they had
> just read the book. Think of books you might use with your students.
> What could you do to give them a greater understanding of the
> context of the book?

To help students engineer for a historical context (specifically, 1630–1730 in colonial New England), their teacher, Julia, read aloud *If You Lived in Colonial Times* by Ann McGovern as part of an integrated social studies and English language arts (ELA) unit. During the unit, the students visited a historic village and spent time discussing, writing about, and imagining what life would have been like for colonists back then. Julia, excited by students' interest, decided to do a Novel Engineering activity and posed this question to her students: As engineers, what could we design that would have made their lives easier? Julia told her students that they would have three class periods and access to "found" materials (e.g., recyclables such as flat cardboard, cardboard rolls, paper, containers, and cloth).

Some of the designs students envisioned were functional—students could test them in class—and some were not. As the activity unfolded, we noticed that students had varying interpretations of the task and their objectives. Some assumed their designs were constrained by the tools and resources that were readily available to people during the colonial era, whereas others assumed that they were designing in the present day (and could magically transport their designs back in time).

✔ **Exercise:** As you read about the group of students building the water filter, look for moments you would classify as productive beginnings of engineering.

An Idea for a Water Filter

Three students—Colin, Jonah, and Brayden—decided to design and construct a water filter (see Figure 6.1, p. 112). From the available materials, they collected three small jars, cotton balls, terry cloth towels, toilet paper rolls, cardboard boxes, and tape. Within the first few minutes of their design work, they had spread everything out and began planning their design. In the following excerpts, we

Figure 6.1: Water filter designed as part of *If You Lived in Colonial Times*

show times when the students struggled to agree on the design objectives, and we highlight aspects of their work that are productive for engineering, such as

- defining the problem,
- identifying criteria,
- conducting and analyzing tests, and
- evaluating their design based on criteria.

1. **Colin:** We're trying to purify water. People in colonial times didn't have much clean water to drink.

2. **Jonah:** So, anyway, we're gonna put a coffee filter in one of the pipes [*refers to paper towel tube*] and when we put water, contaminated water, in there [*points to paper towel tube*], all the gunk and stuff will stay on the filter. And all the water will go into here [*refers to cardboard base*], and we're putting tinfoil around the pipes [*refers to paper towel tube*] so they won't leak.

> ✔ **Exercise:** Part of helping students is anticipating what they will do and what problems they may come across. What questions would you have for the students in this group? These questions should serve a dual purpose. They help you understand students' ideas and help them think proactively about the project and possible difficulties.

The three students defined a problem that people in colonial times faced: They did not have access to clean water. They then explored options for designing a filtering unit colonists could use. In this early description, the students seemed focused on getting their design to function. They considered where the "contaminated water" would enter, how the "gunk" would be cleaned out, and how to prevent leaks.

As we talked to the students, we became interested in understanding what they thought the task was about—their design objectives and how they were going to evaluate their designs. For instance, did they assume that this exact design was meant to work in colonial times, or were they building a prototype for the purpose of understanding and communicating an idea? Jonah's description of their design suggests plausibility of both interpretations. He described the components in terms of both the physical materials (e.g., cardboard tubes and tinfoil) and what he imagines they represent in colonial times (e.g., pipes).

In this early phase of the project, the students seemed to find a creative balance in designing. They considered problems and solutions, and they converged on a solution that would solve a problem for people living during colonial times— one they could physically construct, test, and evaluate in their classroom.

Identifying Criteria

As the boys continued to work on their design, they relocated to the classroom floor, positioning themselves in a circle around three small jars filled with dirty water (see Figure 6.2). They were surrounded by scraps of materials they were using to construct their water filter, such as cardboard box cut-outs, paper towel and toilet paper rolls, paper, cotton balls, coffee filters, rubber bands, tape, and glue. As they discussed their solution, Kathy, a volunteer, curiously asked the boys about their work.

1. **Jonah:**   So, the only problem with it is it might take a long time for the water to drip through the filter into here, but—

2. **Colin:**   That's why we didn't use the cloth I brought.

3. **Kathy:**   Why did you not choose the cloth?

4. **Colin:**   Well, the cloth didn't work. None of the water went through it. We haven't tested the lighter material yet. If it doesn't work ...

5. **Kathy:**   How would you test it?

6. **Colin:**   Well, I'd test it by running water through it on the sink.

7. **Kathy:**   Oh. How fast does it need to go?

8. **Colin:**   Well, we need to get enough water into it for it to go—

9. **Jonah:**   [*overlaps*] Um, maybe like, a cup every 20 minutes or so?

While working on their design, the students iteratively tested their design and identified evaluation criteria. They decided not to use cloth as a filter because it does not allow the water to pass through at a fast enough rate. They therefore planned to test other materials to evaluate filters.

When Kathy prompted them to define a target for water flow rate, Jonah estimated 20 minutes per cup of water. His estimate suggests that he was considering both the possible flow rate of their prototype and the amount of water the people of colonial times would need.

✔ **Reflection:** What criteria have the group identified? How could they test their design to see if it meets the criteria?

Figure 6.2: One iteration of the water filter

Testing and Evaluating

Approximately 15 minutes after the previous discussion, the boys were all intensely focused on watching water drip through each of the filters (soggy cotton balls; wet coffee filters; and a dampened, dirty facecloth) into a jar, and they compared the water cloudiness in each of the jars. This test was designed to address the students' specific design and grew organically out of their desire to optimize that design (see Figure 6.3, p. 116).

1. **Jonah:** So, the towel was good at first, but it tires out easily. Under sustained water, it works good if you just want, like, half a cup. But it's not good if you want to try sustained purifying.

2. **Colin:** But cotton balls, that's what works! Cotton balls have the best results.

3. **Jonah:** Yeah, but I think those cotton balls are a little dirty for … [*pauses and looks pensively up to the ceiling with his hand on his chin*] Wait, uh, guys, can we step back a second? They didn't have cotton balls in colonial times.

Figure 6.3: Testing filters

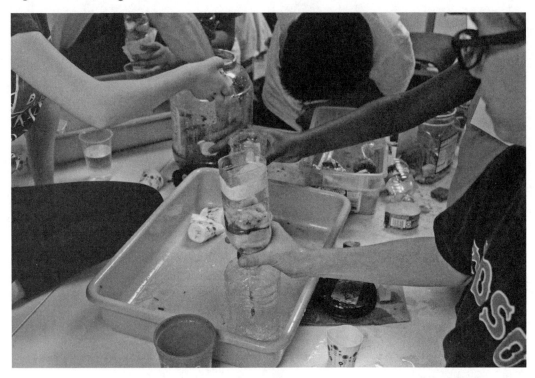

4. **Julia:**   Well, what would they have used instead, do you think?

5. **Colin:**   Probably wool.

6. **Jonah:**   Yeah, but wouldn't that be sort of like the towel or cloth?

7. **Julia:**   Well, I think wool could be a similar type of material, right? I think we could say for now that we could *consider* it wool.

In the first part of this conversation, Jonah described the pros and cons of using a towel as the main filter. He said that it was "good at first" but "tires out easily," and it wouldn't be effective for sustained flow. Colin then added that cotton balls had the best results. Jonah, however, disagreed. He began to argue that cotton balls would get too dirty, but he then had another realization: Cotton balls may have not existed in the colonial era.

When Jonah suddenly realized that they were using materials the colonial people did not have access to, he was unsure how to move forward with their design. For him, it didn't make sense to use materials in the classroom that weren't authentic materials from the colonial setting. Julia swiftly helped him overcome this conflict. She suggested that, for the purpose of their Novel Engineering project, they could use a similar type of material to wool (the setting-appropriate material) that students have in the classroom. In doing so, she helped the students clarify their objectives for the task and reconcile the dual objectives of making a design that is functional both in their classroom and for people during the colonial era. Julia suggested that their prototype could represent a tool that might have been used, but that it does not necessarily need to adhere to the limitations of the colonial era.

> ✔ **Reflection:** The setting of a text can influence the materials students want to use. Sometimes, teachers let students use modern materials to represent ones that are no longer available. Other times, teachers require students to use the same materials the characters would have access to. It often depends on the specific setting of the book and the constraints it places on the problem and solution. Teachers typically let students negotiate this choice themselves. How would you handle this?

Competing Objectives

On the third and final day of the project, the students worked on finishing their engineering design solutions. In the following excerpt, tensions rise between Colin and Jonah as they argue about the main problem.

1. **Colin:**   Well, the coffee filter didn't work. It let the dirt in.

2. **Jonah:**   We can't fix our project. I mean, it's not just that the cloth won't work. If the cloth won't work, that project doesn't work. Even with the coffee filter and the cloth through the bottom, there'd still be a huge puddle.

3. **Colin:**   We could try—we could at least try this. It was the other cloth that absorbed it all. This one is a lot thinner.

4. **Jonah:**   [*frustrated*] It doesn't work enough, though. It leaks, there's huge puddles everywhere, and from the bottom, we don't have anything to stop it. Tinfoil does *not* stop it, and tape does *not* stop it. The only thing that would stop leaks is a lead pipe or something.

With limited time, the boys ended up arguing about how to optimize their design. However, they had different ideas about what their objectives were for the activity. For Colin, ensuring a functional filtering component was the priority. He accepted the limitations of the classroom environment and was prepared to show how their design would function for people of the colonial era. For Jonah, though, the bigger problem was that they couldn't make a functional design in their classroom (at least without leaking) because they didn't have the right materials. He wanted to have a prototype that was built from material the colonial people would have really used, such as wool and lead piping. Eventually, Julia helped them resolve their argument by reminding them that they were designing prototypes and that the people of colonial times may have used similar designs but with different materials.

> Students should be given the freedom to follow their ideas and work through disagreements themselves, but there is a point at which students' disagreements or unreachable project goals are detrimental to moving forward. What strategies do you already use to support your students that would also be useful during Novel Engineering?

Wrap-Up

The students in this group, like many who are engaging in engineering design for the first time, had their share of ups and downs. However, many of their actions and discussions were the productive beginnings of engineering design:

- They defined a problem (lack of clean water).
- They came up with a design to solve that problem (a water filter).
- They iteratively tested different filtering materials to optimize the functionality of their design.
- They defined their own evaluation criteria based on the needs of their clients (colonists).

The students also faced some challenges, specifically, managing the competing objectives of their task. One student remained focused, ensuring the design would function in the classroom setting, whereas another wanted to maintain fidelity to the constraints of the colonial setting. Their teacher played a critical role in helping them overcome these challenges. She helped them understand that engineering constraints existed both in the classroom and for the colonial people. By demonstrating to students that they could make assumptions for their project ("we could consider it wool"), she helped her students overcome their disagreement and continue to make progress on their design.

In successive Novel Engineering units, both with this class and in her other classes, Julia began adding structure to the design task to help students frame problems and navigate open-endedness. For example, when introducing the tasks, Julia modeled reflective questions so students could ask themselves similar questions when embarking on a design. When considering the story setting and characters' needs, she asked, "What would this character want or need in a solution?"

Similarly, when addressing the goals and limitations of the classroom, she asked, "What can we build using the materials available here?" and "Will I be able to test this design to show that it works?" These questions, she found, helped students overcome any difficulty in reconciling the dual objectives so they could design and build a solution that would both help the story characters and be testable in their classroom environment.

Safety Notes

1. Wear safety goggles or glasses with side shields during the setup, hands-on, and takedown segments of the activity.
2. Use caution when using hand tools that can cut or puncture skin.
3. Use only GFI-protected circuits when using electrical equipment, and keep away from water sources to prevent shock.
4. Secure loose clothing, remove loose jewelry, wear closed-toe shoes, and tie back long hair.
5. Wash your hands with soap and water immediately after completing this activity.

Book Resources

America's Champion Swimmer: Gertrude Ederle; Adler, D. A.; Age Range: 4–7; Lexile Level: 800L

If You Lived in Colonial Times; McGovern, A.; Age Range: 7–10; Lexile Level: 590L

Keeping a Shelter Cool Under the Hot Sun

Chapter 7

> As you read this chapter, think about the similarities and differences between science and engineering and how the goals of individual learners may vary depending on their design solution.

Although Novel Engineering focuses on the integration of engineering and literacy, there are also opportunities in Novel Engineering units to connect to other school subjects. We saw this in Chapter 6 when it was used as part of the social studies curriculum. In many Novel Engineering units, just as in professional engineering, students draw on their scientific ideas about the natural world as they scope problems, design, and test solutions. In this chapter, we explore how this dynamic in Novel Engineering units provides opportunities and challenges for teaching and learning science.

There are plenty of other K–12 curricula and units that integrate engineering and science. Just search for "design-based science" or "teaching science through engineering," and you will find many different links to interdisciplinary projects and lessons. A typical pattern among these lessons is that students learn science concepts or mathematical tools and then apply them to solve a design challenge. Other projects might start with an engineering problem that motivates students to learn new science content or practices.

In both approaches, the engineering projects are usually more constrained than in Novel Engineering units. Students are given a more structured problem with a narrower range of possible solutions to ensure that students use (or discover) the scientific or mathematical concept. By contrast, Novel Engineering units open up a wider range of possibilities to frame problems and design and test solutions, which motivates us to think flexibly about the relationship between science and engineering in our teaching.

To help us explore new interdisciplinary possibilities in Novel Engineering units, we turn to *A Framework for K–12 Science Education: Practices, Crosscutting Concepts, and Core Ideas* (NRC 2012) and *Next Generation Science Standards* (NGSS Lead States 2013). Although not all states have adopted *NGSS*, these documents are shaping science education nationwide. In addition to the inclusion of engineering in K–12 science, both of these sources emphasize the importance of students learning the practice of science—that is, how scientists construct knowledge about the natural world.

Therefore, in addition to learning core concepts and ideas, students should learn how to engage in doing science, such as how to articulate questions, design experiments, and develop and refine models to explain phenomena. This shift in science education means that teachers need to help students build and refine their ideas about how to investigate the world alongside their ideas about specific scientific concepts. This shift also requires teachers to include more open-ended projects in their curricula so students have opportunities to engage in science.

In this chapter's case study, we consider how teachers might support students' learning both engineering and science in the context of a fifth-grade Novel Engineering unit. Students in this unit were tasked with solving a problem from *The Swiss Family Robinson* by Johann David Wyss. Specifically, students were presented with the problem of how to keep a structure cool under the sun. This problem invited students to draw on their ideas of heat transfer. Our goals for this case study are to show

- how this Novel Engineering unit surfaced students' science ideas, which could serve as productive seeds for science units on heat and energy transfer;
- that there is productive overlap between the processes of scientific experimentation and of building, testing, and refining engineering designs; and
- that the intersection of science and engineering involves balancing different goals for students' learning.

✔ **Exercise:** Due to time constraints, the teachers in this case study did most of the problem scoping for their students. How do you think this affected students' learning?

Background

This case study focuses on fifth-grade engineering in a rural school district in Massachusetts. Although this Novel Engineering unit was the first one taught in fifth grade that year, several teachers at the school had been participating in the Novel Engineering research project, so many students had prior experience with the approach in earlier grades. It's also worth noting that the Massachusetts fourth-grade science standards include topics around matter and energy, so these students had likely already encountered ideas about the movement, transfer, and conversion of energy in different physical contexts.

For the unit depicted in this chapter, the teachers decided to use two consecutive field trip days as an opportunity to split students into two groups. One group would go on the field trip while the other group completed a Novel Engineering unit; the next day the two groups would switch. This intense, one-day structure was a creative way to fit a Novel Engineering unit into a packed curriculum.

To start the unit, the teachers read an excerpt from *The Swiss Family Robinson*. In the text, the shipwrecked family builds a shelter on the beach, only to find that their new home becomes swelteringly hot during the day. The teachers challenged students to build a prototype shelter that would stay cool under the hot sun. They provided found materials—such as shoe boxes, cotton balls, paper towel rolls, and aluminum foil—and allowed students to get sticks and leaves from the field by the school. The teachers also devised a way for students to test their prototypes using a small cup of water, thermometer, and heat lamp. Given the time constraints, this unit was more structured than most Novel Engineering units—the teachers did the initial problem scoping and devised ways for students to test the designs. The students still had flexibility, however, in how they interpreted constraints and criteria, what they designed and built, and how they justified and analyzed features of their prototypes.

> **Jess's Reflection**
>
> *At the time of the case study, I was a researcher for the Novel Engineering project that supported classroom teachers with this unit. My jobs included setting up video cameras, assisting with materials distribution, and interacting with students. I had been a frequent visitor with this group of students, having supported multiple Novel Engineering units in their fourth-grade classroom.*
>
> *I chose to record one pair of girls I remembered from the previous school year. In their first Novel Engineering unit, Caroline and Amelia worked together to design a dog pen for the characters in the book* Shiloh *by Phyllis Reynolds Naylor. I recalled that they had engaged in extensive brainstorming around the features of the pen and made detailed drawings of their design. As a result, they spent too much time planning. When they finally started building, their classmates had taken many of the materials they wanted, and they ran out of time to finish their prototype. When they presented their unfinished project to their classmates, they were clearly upset and discouraged. I thought they might be an interesting group to record to see how their sense of the design process may have changed.*

Group 1: Planning and Building the Prototype

To start this unit, the teacher, Jess, used an overhead projector to display an excerpt of *The Swiss Family Robinson*. Before she read the text aloud, she asked students to listen and figure out where the characters were located and what problems they were facing. The passage included a quote from the mother character who complained about the "oppressive" heat inside their tent constructed on a "bare rocky spot." The mother further described how the shelter was hotter than the open shore and how she longed to move it somewhere that would be protected from the sun—perhaps into a nearby grove of trees.

> ✔ **Reflection:** What ideas do you think students might already have about the flow and transfer of heat?

While reading the text, Jess stopped at different points to ask students to describe what the characters were feeling, what problems they were facing, and details about the setting. She also asked what students thought particular words or phrases might mean. These pedagogical moves followed their existing classroom literacy practices. The teacher highlighted the problem of building a shelter for the shipwrecked family that would stay cool—setting up that problem as one the students would work together to solve. After discussing the problem, the students grouped themselves into teams of two to four. The design teams were then prompted to work together to draw a plan and build a working prototype.

As the students started drawing their plans, Jess reminded them how they were going to test their prototypes. Their structures had to be big enough to hold the cup of water, and they had to use the materials available in the classroom. Those materials were already set up on a table, and students could go and get whatever they needed throughout the day.

> ✔ **Exercise:** As you read the following excerpts, pay attention to students' ideas about how heat influences their design.

Building the Shelter

When Jess came up to the girls during this project, she was curious to see how they were doing both in terms of their attitudes and with respect to what they had taken away from their earlier experiences.

1. **Jess:**   How's it going, girls?

2. **Caroline:** Good.

3. **Jess:**   Yeah? Did you get all the materials you wanted this time around?

4. **Amelia:**   Yeah.

5. **Jess:**   Yeah?

6. **Caroline:** We haven't had to go back once!

7. **Jess:**   Haven't had to go back once, well that's good. So, what's your design? Can you tell me about it?

8. **Caroline:** Felt keeps you very warm, and it reflects heat as well. Felt also reflects back cold. And I remember that cold air is less thick than warm air.

9. **Jess:**   Cold air is less thick than warm air?

10. **Caroline:** So we have a little chimney for the cold air to come in, but it'll be clogged up halfway with this stuff that's spread out so more cold air would come in than warm air.

Jess first noticed that the girls seemed to be in much better spirits than they were during their last project. They were working collaboratively and had already started building. Caroline seemed proud that they not only had what they needed but also had gotten enough materials that they didn't have to go back for more.

When Jess asked about their design, Caroline started talking about the ideas that motivated their design. She quickly outlined several ideas and, given the noise in the room, Jess remembers having a hard time keeping track of them all. The two ideas she did hear were about the reflective properties of felt and the "thickness" of cold and warm air. When she was asked what she meant by cold air being "less thick," Caroline described how they were using that idea for their design. They were going to spread out cotton balls in the chimney to act as a filter that would only allow in cold air. Jess remembers being confused about how Caroline was thinking of the movement of hot and cold air in her design, so asked for more details.

1. **Jess:**   This is a really cool idea, and I want to make sure I understand it. Heat rises, so that's why you have the chimney on top? Hot air will go out this way, and cold air will come in that way?

2. **Caroline:**   Well, cold goes down and heat goes up.

3. **Jess:**   Cold goes down, heat goes up. So, this is a way to have a little path for the air to go [*gestures*]. Why do you have the cotton stuff on top of here?

4. **Caroline:**   That's for insulation, just to make sure the air doesn't get through. Because that's what you'd have to have if this was made out of sticks, like it's representing.

5. **Jess:**   This is sticks, not—

6. **Caroline:**   This is supposed to be like tons of sticks and stuff.

The students then introduced another idea underlying their design: heat rises. Based on their gestures, Jess connected their idea to the chimney feature of their design. She also noted that they had more cotton on top of the shoe box. Though Caroline seemed to claim the cotton would differentially filter hot and cold air, she later called it insulation, which would prevent all air from going through the holes in the shoe box. She then referred to another aspect of their design—even though they were using a shoe box for their prototype, they modified it so it would reflect the features of an actual structure to be built in the novel's setting.

1. **Jess:**   Okay. You were saying something about where the felt is going to go. It's on the bottom?

2. **Amelia:** There's some on the bottom. We might do it on the side too, so it's harder for the heat [to come through].

3. **Caroline:** We also put some felt on our door so the heat wouldn't get through the cracks. We don't want the cold air since it's less dense to get out the cracks. And we want the hot air to go up there.

4. **Jess:** You said cold air is less dense, so—

5. **Caroline:** So it can fit through smaller things.

6. **Jess:** It can fit through smaller things. Where do you use that idea?

7. **Caroline:** Well, there are lots of things we've learned since we were little kids, about how air—

8. **Amelia:** We are going to put holes in the roof, holes. But we'll have this here, so it'll be harder for the heat to go through that.

9. **Caroline:** Yeah, to come in and then the cold air can go down. So cold air sinks, and hot air rises.

10. **Jess:** Gotcha. So you have two ideas that you're using: cold air sinks, hot air rises, and cold air is less dense so it can fit through smaller things. Awesome. And then you said something about the heat reflecting both? Or the felt reflecting both?

11. **Caroline:** Well, the felt is supposed to keep the heat down, so heat will be down instead of up. That's what felt blankets do. They keep the heat down on you.

Felt played a key role in their design. They had it lining the bottom of their shoe box, and Caroline talked about also putting it on the door to keep the heat and cold from going through the cracks. These two different uses were connected to different scientific ideas. The felt on the door covered the cracks, which Caroline claimed would keep the heat from coming in and the cold air from leaving. At this point, she seemed to pose another scientific idea: cold air is less dense, "so it can fit through smaller things." Jess remembers not understanding what she meant by "dense." Did this idea contradict her earlier idea that hot air rises? Was this connected to her earlier idea that cold air was "less thick" than hot air?

Caroline also talked about felt reflecting heat and cold. Jess didn't understand why they had felt on the bottom of the shoe box. They would be testing their design with a heat lamp held on top of their prototype, so it wasn't clear what role that felt would play in their design. Jess thought one possibility was that perhaps it was there to represent how their prototype would be used on a beach and how the felt would reflect heat from the hot sand.

Jess's Reflection

Caroline and Amelia had such a negative experience in their first project that I was thrilled to see them having a more positive experience this time and feeling capable in using their ideas and building an effective design.

Furthermore, they seemed to have a better sense of the design process than they did during the first project. They spent some time brainstorming and drawing their design ideas, but they also made sure they had enough time to gather materials and realize their ideas. They also referred to the fact that their project was "representing" a solution, offering evidence that they saw it as a prototype for a structure that could be built and used in the setting of the book.

I was also encouraged by how they were making principled decisions about their design based on their understandings of the real world. They talked about each feature of their design in relation to a scientific idea about heat, including how it moves and interacts with other materials.

There were clearly productive aspects of their engineering, but I also had concerns. There seemed to be multiple features to their design functionality: (1) a chimney to channel hot air, (2) materials that could filter hot and cold air, and (3) heat-reflecting felt. However, some of these features seemed to conflict. For example, if the cotton was designed to prevent hot air from passing through, then wouldn't the cotton in the chimney prevent hot air from escaping? Furthermore, since they were implementing these different features all together, it didn't seem like they would be able to identify the features of their design that were successful.

They didn't seem to be thinking systematically about testing and refining their solution. Alongside my considerations of their engineering design, I was also curious about their scientific ideas. I wanted to know more about what they meant by cold air being less thick or dense than hot air. I can imagine that they were drawing on their experiences from hot, humid summers in the Northeast or being in the cold, "thin" air in the mountains.

Caroline also claimed that heat rises and cold sinks. How did that mesh with her other idea about cold air being thinner and less dense? Lastly, I appreciated that they were thinking about how heat interacts with objects, such as felt reflecting heat. These ideas seemed to have some productive seeds for concepts in thermodynamics: heat can flow from one object to another, and objects can prevent this flow. They also showed the beginnings of mechanistic reasoning, namely in Caroline's example of how a blanket traps and reflects heat.

However, as a science educator, I wanted them to better articulate how they were thinking about heat flow—why hot or cold air would slip through the cracks of their design and what happens when an object reflects heat. I also wanted to help them refine their ideas to be more in line with science content objectives around heat transfer.

> ✔ **Reflection:** How might these different responses address different goals in science and engineering? If you were the teacher, what else might you think about as you decide what to do next?

What Next?

Jess remembers feeling conflicted about what to do next. She could see pursuing multiple avenues to support the students' design, such as encouraging them to independently test the different features of their design or asking them to draw how each worked. She could also imagine supporting their scientific work, perhaps by asking them to write about how heat flows or asking them to draw a diagram showing how heat moves through their design.

However, if Jess interrupted their building, she might keep them from finishing their design and experiencing the full design process—something they didn't get to do in their previous project. She could also let them continue building and check back after they finished, but then there might not be enough time to return to these issues. In the end, she hoped that letting them continue to build might leave opportunities for them to discuss and refine their ideas—about both engineering design and science—and would help them see themselves as capable of doing engineering.

Group 2: Testing and Revising the Prototype

To help students test their designs, one of the fifth-grade teachers set up a station with a heat lamp, a cup of tap water, and a thermometer. As students prepared to test their shelters, they placed a thermometer in the cup and measured the temperature of the water. They then placed the cup and thermometer inside their shelter. The teacher held the heat lamp above the prototypes for two minutes so students could check the final temperature of the water.

As students headed back to their desks to wrap up, Jess heard a student, Laura, complaining to another student, Rebecca, about the possibility of having to write something about their design, which she found "annoying." When Jess asked if she had enjoyed doing the engineering project, Laura was much more enthusiastic.

1. **Jess:** What did you like about it?
2. **Laura:** I like figuring out, like, what we should do.
3. **Jess:** Yeah.
4. **Laura:** And finding out what would work and what wouldn't.

5. **Jess:** Yeah? What did you find out that worked?

6. **Laura:** Tinfoil.

7. **Rebecca:** Oh, I liked when you're not the first person to test. We got to see that the duct tape was melting and the tinfoil was keeping it cool. So I liked that since we were some of the last people to go, we got to see those mistakes.

8. **Jess:** Yeah?

9. **Rebecca:** And then we tested ours and then we saw what we did wrong so we added and made stuff different.

10. **Jess:** You did it twice? You did two tests?

11. **Rebecca:** Yeah, we did two tests. The first one was our regular thing, and the second one had—

12. **Laura:** More aluminum foil.

13. **Rebecca:** Yeah, more aluminum foil.

14. **Jess:** Why does the aluminum foil work?

15. **Laura:** Because it, uh ...

16. **Rebecca:** It reflects.

17. **Laura:** It reflects.

18. **Jess:** What does that mean?

19. **Laura:** It bounces off.

20. **Rebecca:** The heat's bouncing off the tinfoil.

Jess noticed immediately how Laura smiled and spoke at a quicker pace when talking about the engineering. Jess also was struck that she enjoyed the process of "figuring out" what to do and what "worked." At this point, Rebecca joined in and described how they watched other group's tests and learned about the integrity of materials under heat and how well they performed. They then used that knowledge to interpret what went wrong when they tested their own design and redesigned their prototype by adding more aluminum foil.

Jess's Reflection

I had not interacted with these girls earlier, but when I happened to hear Laura complain about possibly having to write, I was curious to find out about her impression of the project. I was encouraged by her noticeable change in affect, but what stood out to me was that she was most excited by figuring stuff out. To me, it seemed like she viewed her work in this project as a process of learning. Both Laura and Rebecca learned from building their design, testing to see how their and others' designs worked, reflecting on the results of their design, and then revising based on what they had learned.

In many ways, this is similar to how we want students to view learning in science. We want students to build knowledge about phenomena, test that knowledge by designing experiments or comparing it to other knowledge, and then reflecting on the results to revise their understandings. I could see connections between engineering and science in how Laura and Rebecca viewed their work as engaging in figuring things out.

I also noticed that the girls were not just attending to what affected the temperature in their tests; they were also considering other aspects of their design. For instance, they pointed out issues related to structural integrity by observing that the duct tape melted under the heat lamp. They also thought about how the materials worked to keep the structure cool, tapping into their ideas about heat transfer. Laura and Rebecca thought that the aluminum foil was a good material because it reflected heat, which seems like a similar idea to the one that Caroline and Amelia expressed about the felt. This made me wonder if the two pairs of girls were thinking the same way about heat being reflected. For instance, would they say that felt and aluminum reflect heat in the same way? What materials do they think do not reflect heat? I also wondered what they might say if asked to reconcile their experiences of metal being a good conductor of heat with their ideas about how aluminum kept the structure cool.

I was also concerned about the testing set-up and how the other teachers and students were treating the results as unproblematic. From my perspective, there were several experimental design flaws or places for measurement error. For example, the lamp was not always held a consistent distance away from students' prototypes, the same cup of water was used in back-to-back tests (so for later tests, the water started at a much higher temperature), the thermometer was not sensitive enough to measure small changes in temperature, and so on. Indeed, one group of students found that the temperature of the water went down by four degrees during their test (which is not scientifically possible!).

(continued)

(continued)

However, part of learning science and engineering is learning not only to discover ideas but also to evaluate them. This test provided opportunities for students to critique the set-up and consider how errors might affect their results. I did not see such conversations take place during the short time available for this unit, but if there had been more time, I could easily imagine a teacher asking students to think about where there might be measurement errors or other complicating factors in their testing procedures.

In engineering, a discussion would center on how these experimental design flaws might affect our understanding of how a prototype performed. In science, a discussion would center on the impact those flaws might have on our understanding of heat flow. For both disciplinary goals, students should have opportunities to iterate and refine not only their engineering designs but also their testing procedures.

Looking Forward: Possibilities for Integrating Engineering and Science in Novel Engineering Units

The following day, the teachers modified their lesson plan to include a mid-design share-out. This discussion allowed them to hear students' ideas as they were preparing to test their prototypes. During this share-out, students voiced familiar science ideas about how hot air rises and how heat reflects off certain objects. They also heard new ideas about how objects interact with air at different temperatures. For example, one group came up with an idea to use a straw with a pipe cleaner inside as a chimney, suggesting that the pipe cleaner would "capture" the cold air and keep it inside the box. Eliciting students' ideas about how their designs function helped teachers recognize how those ideas changed as students continued to manipulate materials, discuss with one another, and reflect on the results of testing.

Given the short time allotted for this project, there was not enough time to pursue students' scientific ideas further. However, there are several possibilities for what teachers might do next if they have time. For instance, teachers could collect and make a list of students' diverse ideas for how heat moves through their designs and what can stop that movement. Students could work to develop representations of these mechanisms, drawing or even creating stop-motion animations (e.g., with HUE Animation or a similar app). Students could also read about the different types of heat transfer, such as conduction, convection, and radiation, and then make connections between these and the ideas they generated in their designs.

> ✔ **Reflection:** If you have already done an engineering activity with your students, what overlaps or tensions have you seen between engineering design and science? When there are tensions between science and engineering goals, when might you emphasize students' learning in engineering? When might you emphasize their learning in science?

This case study illustrates that integrated engineering and science units can be more than just a final assignment for students to demonstrate their understanding of science concepts. We see Novel Engineering as a jumping-off point for students' science explorations. In this unit, we discovered a rich terrain of students' ideas—not just about how to design a structure on a beach but also about the phenomena of heat and energy. This presents an exciting opportunity for teachers since engineering design provides access to students' knowledge and about how students think the world works.

Novel Engineering can also help students learn science concepts as they make connections between those concepts and their prior knowledge. Drawing on this knowledge and helping students elaborate, investigate, and refine their scientific ideas provides a way to make science learning more meaningful and authentic and to value students' diverse ideas and perspectives about scientific phenomena.

Safety Notes

1. Wear safety goggles or glasses with side shields during the setup, hands-on, and takedown segments of the activity.
2. Use caution when using hand tools that can cut or puncture skin.
3. Use caution when working with heat sources (e.g., lamp) that can cause skin burns or electric shock.
4. Use only GFI-protected circuits when using electrical equipment, and keep away from water sources to prevent shock.
5. Secure loose clothing, remove loose jewelry, wear closed-toe shoes, and tie back long hair.
6. Wash your hands with soap and water immediately after completing this activity.

References

National Research Council (NRC). 2012. *A framework for K–12 science education: Practices, crosscutting concepts, and core ideas.* Washington, DC: National Academies Press.

NGSS Lead States. 2013. *Next Generation Science Standards: For states, by states.* Washington, DC: National Academies Press. *www.nextgenscience.org/next-generation-science-standards.*

Website

HUE Animation: *https://huehd.com*

Book Resources

Shiloh; Naylor, P. R.; Age Range: 8–12; Lexile Level: 890L

The Swiss Family Robinson; Wyss, J. D.; Age Range: 8–12; Lexile Level: 630L

Engineering With Empathy and Compassion

Engineering is often seen as a profession for those who excel at math and science, which may limit the diversity of people who pursue it as a career. As you read this chapter, think about other types of interests and skills that engineers use. Also consider how empathy plays a role in students' designs.

In this chapter's case study, we look at a fifth-grade classroom reading *El Deafo* by Cece Bell as part of a Novel Engineering unit. Although engineering is traditionally considered a technocentric discipline focused on the application of the so-called hard sciences, we've seen that engineering is also a discipline charged with developing solutions to human problems that must work in complex socio-cultural contexts. This case study explores the idea of design empathy as central to engineering. The main character in *El Deafo* is Cece, a deaf girl who faces many challenges with hearing, communicating, making friends, and fitting in. In response, the fifth-grade students explored a number of solutions meant to help Cece amplify sound in an inconspicuous way. This chapter includes incidents taken from the classroom interactions where the students adopted a compassionate design approach to their engineering work.

Engineering is often described as a profession focused on the development of new technologies through the application of mathematics and science knowledge. To become an engineer, many people think you have to love mathematics and science and excel in these subjects in school. One issue with this perspective is that it can limit the types of people who want to pursue engineering. It is true that many engineers fit into this mold, but there are many other skills and interests that relate to becoming an engineer.

In developing a solution for a person or group of people, engineers are challenged to understand their intended users. Human-centered design (HCD) is an approach that many engineers are taught and then practice when they work as engineers. The principles of HCD relate to understanding the human context(s) for which a solution is developed. In the problem-scoping phase of the engineering design process, engineers may conduct user focus groups, collect survey data from potential users, or rely on their knowledge of people and their behaviors and cultures, which is knowledge that comes from social science subjects such as psychology, economics, or anthropology (Hynes and Swenson 2013). This is much of what makes the work of the engineer difficult; people and their preferences, needs, and behaviors are much more unpredictable than the mathematical and scientific principles upon which engineers draw.

Delving a bit deeper into the HCD approach, compassionate design is an approach that intentionally considers the users' emotional well-being as it is related to security, dignity, and empowerment (Seshadri et al. 2019). This is not necessarily an appropriate approach for all designed solutions, but it can be particularly useful when designers or engineers develop solutions around medical devices, for people in developing countries, or in contexts where users experience a lack of security, dignity, or empowerment. As will become evident, compassionate design is appropriate for students' work with *El Deafo*, since the main character at times experiences a lack of dignity—her loss of hearing leaves her feeling marginalized.

> ✔ **Reflection:** Are there any books you have read with your students that you feel would meet the criteria for compassionate design? What types of problems could students tackle, what additional reading would you provide, or what research would you ask them to complete so they could better understand the characters' problems?

Given the humanistic nature of engineering, it is critical that engineers are prepared with the appropriate knowledge and skills to tackle human-centered problems. The case presented here illustrates how Novel Engineering can provide a meaningful environment for students to explore social and emotional problems people really face and then design solutions with both technical and nontechnical considerations.

This case study begins with third- through fifth-grade teachers at an Indiana elementary school. The teachers participated in a Novel Engineering professional development workshop, discussed the approach, and worked with a researcher/facilitator to identify an appropriate book and plan the activity. Margaret, a fifth-grade teacher, was excited to read *El Deafo* with her class. She liked how the book touched on profound emotional issues related to fitting in at school and making friends, and she thought it was especially relevant to some of her own students.

El Deafo is an autobiographical account of a young girl who grew up deaf. As a graphic novel, the book depicts the protagonist, Cece, as a cartoon rabbit. Throughout the story, Cece faces numerous challenges related to her education, friendships, emotional well-being, and sense of self. The book has some deep character development, and the graphic novel format provides additional details that help students visualize how their solutions fit into Cece's particular context.

The following account shares details from various days during Margaret's Novel Engineering unit.

Day 1

Margaret started the first Novel Engineering discussion of *El Deafo* by asking students to share key words they wrote down after reading the first 50 pages of the book. She explained that these words should be important to understand what the book was about. The student-generated list included the key words seen in Figure 8.1 (p. 138).

> ✔ **Exercise:** As you read this excerpt, identify moments that reflect students' engineering and literacy thinking. What role does paying attention to Cece's well-being play?

Margaret then pursued one key word in particular—*different*. The conversation that follows represents Margaret's back-and-forth with her students regarding how Cece may have experienced *different* and what sorts of challenges or problems being *different* presented.

1. **Margaret:** I'm particularly interested in *different*. Why do you think different is going to really affect Cece?

2. **Student 1:** She's sort of set apart from the rest, in her own little bubble like you were talking about earlier. She's different.

Figure 8.1: Generating a list of key words from *El Deafo*

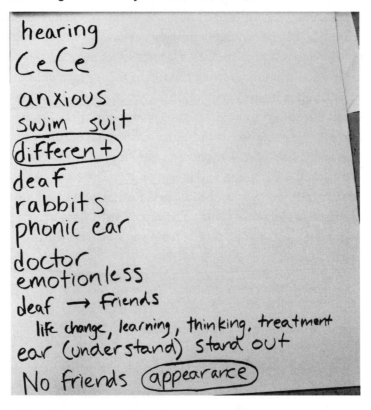

3. **Margaret:** This whole book is a good example of cause and effect. Something happens and then all these other things happen afterward because of that one thing. She had this big problem—she got sick and then became deaf. That was the big problem.

4. **Student 2:** Deaf?

5. **Margaret:** Deaf, but then beyond that, what were some of the other problems that were creeping up because of this big problem she has? Think about that. What problems have you identified that she has because of her deafness? So being deaf is the big problem, but what does it lead to?

6. **Student 3:** No friends.

7. **Margaret:** Okay, it was affecting her and her friends?

8. **Student 1:** Her social life.

9. **Margaret:** That's a good way to put it. What else?

10. **Student 1:**   It changed her way of life.

11. **Margaret:**   This is definitely what some people like to call a life-changing event. This was a life change—a big one.

12. **Student 3:**   It affects her learning.

13. **Margaret:**   Oh boy! She had to go to a whole different school, didn't she? Did her learning get better or worse?

14. **Student 4:**   Worse.

15. **Margaret:**   It was more difficult. That would've been a better way to put it. What else?

16. **Student 3:**   Thinking.

17. **Margaret:**   It definitely changed the way she thought about the world and the way she saw things. What other problems does she have?

18. **Student 3:**   The way people were treating her.

19. **Margaret:**   Her treatment changed; people were probably looking at her in different ways.

20. **Student 2:**   Her hearing.

21. **Margaret:**   Definitely her ability to hear. If you can't hear, you can't—

22. **Student 2:**   See?

23. **Margaret:**   Well, you can see.

24. **Student:**   If you can't hear, you can't listen or hear directions.

25. **Student 2:**   You can't understand.

26. **Student 4:**   And that's why she was wondering why all the kids were getting ice cream and she wasn't.

27. **Margaret:**   Yeah, she's just kind of sitting there and figuring that they didn't ask her yet. What else? What problems have you identified so far? There were some others. Think about what she did or how she felt after she saw the hearing specialists.

28. **Student 4:**   She felt like the hearing aids were going to make her stand out, making her alone.

29. **Margaret:**   She felt like she stood out. Why would she feel that way?

30. **Student 4:**   Because none of the other kids at her new school had to use hearing aids.

31. **Student 2:**   Besides that one class she's in.

This exchange between Margaret and her students highlights the attention they paid to Cece's social and emotional well-being. Margaret's line of questioning helped students take on the perspective of their prospective client, Cece. This is something designers and engineers do as they develop solutions for specific clients or groups of clients. By relating to the experiences of a client or potential user, the designer or engineer is better equipped to make decisions that are in the best interests of those users and to address issues that may not be readily visible or easily identified early in the problem definition stage. Such a conversation is a great first step in helping students identify with the client, but it also relates to literacy learning objectives and helps students develop a thorough understanding of a story through its characters.

This conversation continued to develop as Margaret pushed students to consider a more specific issue that made Cece feel different—namely, her appearance.

1. **Margaret:** How did that affect her? She was feeling like she was really different. Can you think of some different ways she might have felt different besides what she mentioned? This is kind of a small thing but can be big for a girl.

2. **Student 1:** She doesn't have any friends yet.

3. **Margaret:** You know, for children, friends are number one.

4. **Student 2:** She always thought people were looking at her funny because she had things sticking out of her ears.

5. **Margaret:** Okay. I'm going to put that down as her appearance. What about her appearance? What changed about her appearance?

6. **Student 2:** She used to look normal like everyone else, but then she got meningitis and had to get hearing aids. And no one at her new school had hearing aids, so she was just kind of alone.

7. **Margaret:** How would someone know she was wearing hearing aids?

8. **Student 2:** She had cords. [*motions that the cords stretches from her ears to her chest*]

9. **Margaret:** She had cords coming out of her ears. Just think, nobody else has these and you're walking around with these cords. Just think about that. Do you think this would have made you feel funny?

10. **Student 2:** Maybe. Possibly.

11. **Margaret:** Possibly, yeah. Besides the cords, what was something else that made her look different? Something they gave her to wear.

12. **Student 2:** I forgot.

13. **Margaret:** Where did the cords go?

14. **Student 2:** In the pouch. She had this big thing.

15. **Margaret:** In the pouch. She had this thing hanging over her neck. She didn't want everyone looking at that, so what did she do?

16. **Student 1:** She covered it up.

17. **Margaret:** She tried to cover it up by putting her shirt over it. But do you think it was completely invisible?

18. **Student 1:** No.

19. **Margaret:** Right. Her appearance was important, but she had this pouch around her neck and she had to wear that to school.

The conversation continued to develop the socioemotional depth of the book's main character, and it helped students think about how various social and emotional ideas may connect to problems they might solve as engineers. Margaret smoothly transitioned from this conversation to one about possible solutions that might address Cece's problem—her hearing aid device has a negative impact on her appearance, which then affects her self-esteem, ability to make friends, and relationship to her new school.

The students brainstormed ideas to reduce the size and visibility of Cece's hearing aid. They thought about having a small "pea" behind her ear instead of the bulky hearing aids with cords, some sort of invisible hearing aid, and even brain surgery to implant a hearing device. They then shifted to ideas that would integrate the device with a hat (or something Cece could wear on her head) or a device that works with a new hairstyle and would cover up the cords. These conversations were exciting because Margaret had focused her students' attention on thinking deeply about the problem and about solutions that go beyond a technical solution and consider the feelings of their client. Margaret wrapped up the conversation by asking her students to continue reading the story and thinking about how Cece felt different from everyone else. How could they might help her as engineers?

Day 2

For the next class session, Margaret recapped the previous discussion and reviewed the list of problems and potential solutions students discussed. She then shared a few newspaper and magazine clippings of products and devices somewhat related to the problems they had discussed—wireless headphones, a memory storage device for mobile phones, and a music app for phones. She then built on their conversation by asking students to think about a smartphone and all the accessories and extra functions it can have beyond making phone calls.

1. **Margaret:** It occurred to me that a lot of our ideas for making our project to improve her life were pretty serious. Have you thought about Cece's phonic ear or other things she wore? What if she had other accessories that would make her different or special compared to the other kids? She would still have the hearing aid, of course, but can you think of ways it could be adapted that would make it kind of cool?

2. **Student:** Maybe Cece should have it be an earring in her ear.

3. **Margaret:** Great! An earring with the hearing aid. To give you an idea of how many accessories there can be, I thought we would make a web about cell phones, which are so popular. First, talk to someone next to you about all the things cell phones can do. Then you are going to write down five of your best ideas and we'll post them here. [*motions to board*] I'm hoping this will open up your minds and give you ideas for your project.

Margaret proceeded to give her students sticky notes for their ideas and then placed them in a concept web. Margaret was demonstrating that a device designed to solve some primary problem (e.g., making a phone call) could have additional features or functionality that would enable it solve additional problems (e.g., playing music). This was a clever way for her students to see that multiple problems from the long list of problems they had brainstormed could be addressed with a single engineered solution. Margaret then encouraged her students to consider additional features or functions that would address multiple problems Cece faced.

Before having students sketch out their own solution ideas, Margaret asked her students to revisit an incident in the book between Cece and her mother. During this incident, Cece got upset when she learned she was moving to a new school—one that was not specifically for deaf children. Cece got so mad that she envisioned kicking her mother—and then actually kicked her. Margaret asked her students to put themselves in Cece's shoes and think about why she may have acted like that.

This was an excellent opportunity for students to bring empathy and compassion into their thinking, which would eventually factor into how they developed their solutions. The students responded by saying that moving to a new school might be scary for Cece because she would look different, which might make it difficult for her to make new friends and fit in. Margaret then asked students to look at the incident from Cece's mother's perspective. The students shared how it was unfair to the mother and that the mother was just trying to do what was best for Cece.

Margaret created an awareness among her students to consider multiple stakeholders and their perspectives. Whether she approached this activity from an engineering perspective or from a literacy perspective was unimportant. Her approach successfully addressed learning objectives from both perspectives.

> ✔ **Reflection:** What kinds of activities do you already do with your students that help them structure their thinking? How could you modify them to work with a conversation like this?

Day 3

Margaret had her students complete a worksheet where they listed their top three solution ideas. Having students go beyond their first idea and come up with multiple ideas has a number of benefits. First, idea fixation—or sticking to your first idea—is a common problem among beginning designers. Though it is possible your first idea ends up being the best idea, not allowing yourself or your team to continue brainstorming limits the perspectives you consider, the types of solutions you develop, and the participation of teammates whose ideas may not be discussed.

On the third day, Margaret had her students sketch out their ideas and identify materials they needed to prototype their ideas. This is one way to have the sketching process be authentic for students. Sometimes, sketching feels forced since students already know how they want to build their prototype. However, if they are motivated to get specific materials, they become more invested in communicating their ideas through a sketch. This approach models what many engineers do—draft drawings or plans for ideas and then share with others to develop.

Day 4

On the fourth day, the students were given a variety of materials they could use to prototype their solutions. Their solutions ranged from a simple disguise for Cece's hearing aid to a solution that would allow her to watch TV at home to a cure for meningitis. As with many implementations of Novel Engineering in the classroom, some students ventured off to solutions that were not be feasible or realistic given the context. For example, students will not develop scientifically possible solutions if their goal is to create a cure for meningitis.

However, there is a benefit to allowing students to pursue their own ideas, which can then be supplemented by having them conduct additional research. For example, one group thought about how to disguise Cece's hearing aids. Their ideas included hearing aids embedded in sunglasses, a backpack, or a T-shirt and hearing aids disguised as a phone clipped to the user's ear. Given the conversation Margaret had with her students, it became clear that students were able to address multiple problems with their solutions.

1. **Margaret:** So, what are you prototyping?

2. **Student:** Cece always complained that the phonic ear was really annoying because she always had to wear it here [*motions to her chest*] and it was really hot.

3. **Margaret:** Okay.

4. **Student:** So we made it into a backpack and the wires wouldn't be as noticeable because it wouldn't be coming here. [*motions to the front of her body*] The cords would be coming from the backpack.

In this example, the students addressed both the issue of Cece's appearance (by hiding the wires and phonic ear in a backpack) and the issue of the pouch being hot and uncomfortable (by moving it into the backpack). Both decisions come from a place of empathy or compassion; the students considered Cece's physical and socioemotional needs. Another team developed an over-the-ear device that would look similar to a wireless phone earpiece, and they added that the device would work both as a hearing aid and as a phone. They explained, "Because if people think you have an advanced phone thing like that, they would think it's cool and wouldn't even know it's a hearing aid." This highlights how they, as engineers, considered the perspectives of people who may be looking at Cece and how the solution might make her look "cool."

> ✔ **Reflection:** What are other ways that students could share their designs in a unit so heavily focused on user-centered design and empathy?

Wrap-Up

It is possible that each implementation of Novel Engineering can take a unique direction. In this case, *El Deafo* provided students with an opportunity to consider the main character's emotional issues and need to use an assistive hearing device. They were able to experience how engineering could solve a problem technically (helping Cece hear better) and solve a problem related to the user's feelings and emotions (helping Cece fit in).

For her part, Margaret did an excellent job intentionally setting this up with the book. In the literacy portions of the lesson, Margaret emphasized how Cece's hearing device affected her appearance. She then had students think about how Cece's appearance made her feel. This led to much of the engineering design thinking in which students developed solutions that disguised the hearing aid in clothing or accessories that might help her look cool.

From human-centered design and compassionate design perspectives, students engaged in the authentic practices of working to understand their client and recognizing the variety of criteria and constraints related to their client's physical and emotional needs and well-being. With this approach, more students make personal connections to the engineering than they would with a less client-centered approach.

Safety Notes

1. Wear safety goggles or glasses with side shields during the setup, hands-on, and takedown segments of the activity.

2. Use caution when using hand tools that can cut or puncture skin.

3. Use only GFI-protected circuits when using electrical equipment, and keep away from water sources to prevent shock.

4. Secure loose clothing, remove loose jewelry, wear closed-toe shoes, and tie back long hair.

5. Wash your hands with soap and water immediately after completing this activity.

References

Hynes, M., and J. Swenson. 2013. The humanistic side of engineering: Considering social science and humanities dimensions of engineering in education and research. *Journal of Pre-College Engineering Education Research (J-PEER)*, 3 (2): 31–42.

National Academy of Engineering. 2008. *Changing the conversation: Messages for improving public understanding of engineering.* Washington, DC: National Academies Press.

Seshadri, P., C. H. Joslyn, M. M. Hynes, and T. Reid. 2019. Compassionate design: considerations that impact the users' dignity, empowerment and sense of security. *Design Science* 5 (e21).

Book Resource

El Deafo; Bell, C.; Age Range: 8–12; Lexile Level: GN420L

Section III

Enacting Novel Engineering

Introducing Novel Engineering and Thinking About Classroom Culture

Chapter 9

This chapter discusses how to introduce Novel Engineering to students. If your students haven't done any engineering yet, it is important to have a talk about what engineering is and what engineers do before starting a Novel Engineering unit. We've included a few introductory activities that will set the stage for students to understand what they will be doing and the classroom culture that will evolve during Novel Engineering.

Introducing Novel Engineering to Students

As with any new activity, students benefit from an introductory discussion to activate prior knowledge, provide context, and describe expectations. Once students have a sense of what engineers do and what engineering involves, they are more apt to jump right into a Novel Engineering project, thinking of themselves as young engineers.

Many teachers introduce the Novel Engineering project and the book they are using for the project at the same time. For example, a teacher might begin with, "We are going to read *Muncha! Muncha! Muncha!* and act as engineers as we read it." If the teacher chooses to discuss engineering at this point, he or she might then open the conversation up by asking students, "Does anyone know what engineering is or what engineers do? We see the output of engineering all around us—cars, bridges, computers, gasoline, clean water, electrical power, wireless communication, and more. Look around this room. What do you think has been built with the help of an engineer?"

These opening questions set the stage for a discussion similar to the one Maggie facilitated when introducing *From the Mixed-Up Files of Mrs. Basil E. Frankweiler* (see Chapter 4). The aim of the discussion is to work toward a shared understanding of what engineering is and what engineers do so you have a touch point when talking about students' work. A good working definition of engineering may include the following ideas:

- Engineers design things to solve problems.
- Engineering design is an iterative process.
- Engineers test their designs many times.
- Understanding a testing failure helps make the design better.
- Engineers typically draw on other kinds of knowledge, such as math and science, when they solve problems.
- Engineers work to create solutions to help specific users.
- Engineering is a collaborative process that requires teamwork.

Students should be reminded of the ways in which the problem solving they already do on a daily basis is similar to the work of engineers, but understanding the engineering design process (EDP) is critical to the activities students do within Novel Engineering. We encourage teachers to let students participate in a short design experience before going into detail about the EDP and its steps. This is done for two reasons. The first is that if students do not have engineering design experience, it might be difficult for them to fully participate in a conversation about the EDP. It easily becomes a lecture with the teacher telling them what they should be doing at each step. If students have experiences to reference, however, they can make connections between what they did and what the teacher is saying, and the interaction turns into a discussion (rather than a lecture). The second reason is that if students are presented with the EDP's steps and see them as the goal of the project, the EDP often lacks fluidity and becomes a checklist of linear steps rather than an interactive process.

If your students have had little experience with engineering, it may be helpful to do one or a series of short (30–60 minute) activities designed to introduce them to specific aspects of engineering design. This may help students understand what they will be doing and what is expected of them. These quick activities can provide some context to start conversations about engineering and make the discussions more equitable. Some students may know someone who is an engineer, whereas other students may have limited or no knowledge about engineering. Making sure each student has at least some knowledge of engineering will help all students contribute to productive conversations. These experiences will also serve as a reference point when students engage in larger, more complex engineering design.

Introductory Engineering Activities to Build Foundational Skills

Introductory activities can be quick engineering tasks that help students understand what they will be doing and what is expected of them during Novel Engineering units. They can also help students build foundational skills. Introductory activities should do the following:

- Provide a place to start conversations about engineering
- Give everyone some experience
- Act as a reference to larger, more complex activities
- Help establish the classroom culture
- Provide students with relatively low-stakes engineering experience

There are many hands-on activities that can provide a first engineering experience to students, help them understand the properties of materials, and clarify expectations of their participation. For example, designing an aluminum foil boat to hold coins is not a very robust activity. However, it will still help students understand the concepts of testing and iteration as well as the idea that engineering involves creating testable, functional solutions. Other simple activities include making a tower or table out of paper and tape or creating a catapult to shoot a marshmallow or ball.

Discussing the Pros and Cons

Students, especially younger ones, often think there is one "best" design solution. This activity, not based on a text, helps students weigh the pros and cons of a design and realize there are many things that contribute to design choices. Choose several versions of one tangible item (e.g., water bottles, pens) and ask students to pick the "best" one from the available choices. It is better if the item is something with which students are familiar and is personally meaningful. We will highlight examples of this activity from three different grades.

In the first example, first graders were given four different pencils and asked to pick the best pencil. As a class, they agreed on four characteristics they would use to judge the pencils: attractiveness, cost, ease of sharpening, and feel (see Figure 9.1, p. 152). Using a worksheet as a guide, each group of four students discussed how the pencils ranked in the four categories (see Appendix D, p. 229 for

Figure 9.1: Pencil comparison in first grade

Grade 1   Pencil Comparison Project

Group Members _____

#2 pencil (regular pencil)	bord enosh broks	good they sterng dot brake	bad they can get small	good the eraseerson the botom	bad they can brak down the mifill
Thick Pencils	It is thatito	they cant chock onit good	to dig to sharpin	no erasers on the dotom bad	it is thick
Mechanical Pencils	has eraser on the dotom good	they can make the lied out	they can split down the mitell	the string can not brake good	they cant chock on the botom good
Novelty Pencils	they are sharp dab	that are hard to sharpin	they look coll good	the paper can pell off dab	good the eraseron the dotome good

a blank worksheet that can be used for any item chosen). They were then asked to name the best pencil for kindergarteners, which sparked another set of conversations. Pencils that were previously favored were abandoned as students thought of the needs of younger students. Finally, each group picked a pencil and argued for their choice to the whole class, which would ultimately recommend one pencil for kindergartens. This exercise helped students consider the needs of others (e.g., thicker pencils are better for smaller hands) and their preferences (e.g., younger students would like pencils with animals on them), and it helped them realize that a design choice is contingent on context—there is often not one definitive "best" in all situations.

A second-grade class followed a similar process. They were told to pick the best folder for their writing workshop. In this case, the teacher said she would buy the one that was chosen by the class for them to use during the rest of the year. The students considered durability, attractiveness, cost, environmental impact (plastic versus cardboard), and number of pockets. After they picked the best folder for the job, each group was then tasked with writing a letter to the teacher, explaining which folder they had picked and why (see Figure 9.2).

Figure 9.2: Letter written by second graders explaining their folder choice

Dear Mrs. Miller

We recommend that we use the following folder for Writer's Workshop:

2 - pocket folder with tab

It is a good folder for Writer's Workshop because

It is cheap. When you fill up the slots theis more room in the middel. It can stand up for a long period of time like for a test. It's light like in weight.

Sincerely,

Likewise, a fifth-grade class was asked to pick the best water bottle among several given choices. Students identified cost, reusability (environmental impact), size, weight, and ease of being carried as factors for consideration. In this case, students were not working to find the best solution for others but were instead trying to convince their class of the best solution for themselves. During their discussions, the fifth graders hypothesized about what type of person would be the best fit for each bottle.

In each of these three situations, students had prior experience with the item and felt personally invested in the outcome. This influenced their discussions and their final choices. With all groups, the teachers allowed time for students to reflect on the exercise and connect it to the work they would be doing in Novel Engineering; the teachers wanted their students to realize there is not one best solution and that an appropriate solution is heavily tied to context and users.

Exploring the Properties of Materials

You can have students explore the properties of materials by giving them a few different materials and asking them to assess and compare the strength and durability of the materials. An sample worksheet you can use to help students keep track of their findings is in Appendix E (p. 230). Students can test the materials either by manipulating and testing them alone or as part of another building activity. Teachers have done this a few different ways, depending on the amount of time they want to spend on the activity. For example, small groups of students (e.g., four) could explore each material, or different groups could work with different materials and then report out to the whole class.

Finding Problems and Brainstorming Solutions

We include this as an introductory activity because you can give students practice finding problems and brainstorming solutions while reading a book even if you are not going to include the book as part of a larger Novel Engineering unit. In this activity, students identify problems and brainstorm solutions, but they don't plan or build the solution. Have students name a few problems they found while reading, or have them keep track of problems as they read. Add their suggestions to an anchor chart as you talk about the book during whole-class discussions.

To give students brainstorming experience, pick one of the problems from the student-generated list and brainstorm some solutions as a group. The following exchange was taken from a longer conversation about the pros and cons of using a wig as a solution to a problem in Sara Pennypacker's *Clementine* (see Chapter 3).

1. **Holly:**   Um, me and Cecelia were thinking of pros and cons too and we—
2. **Teacher:**   Great. For the wig?
3. **Holly:**   Huh?
4. **Teacher:**   For the wig idea?
5. **Holly:**   Yeah for the wig, and one of the pros was that she could still go to school and no one would really notice it, but the bad one would be, what happens if, like, she goes upside down or something or—
6. **Mary:**   On monkey bars.
7. **Teacher:**   I'm going to just put a little extra star next to "it could fall off," depending on what she's doing.
8. **Mary:**   Or if they touch her hair, they would feel it's fake.

9. **Teacher:**   Ah, okay.

10. **Dash:**   Oh yeah. That's a really good one.

11. **Teacher:**   Ah, okay. I'm going to say it feels fake. It feels fake.

12. **Teacher:**   What were you guys saying about how you were going to attach it to her head? You guys were talking about something very interesting there.

13. **Mary:**   Well, I was thinking tape.

14. **Holly:**   But I was thinking glue. But then when some of her hair starts to grow back, she wouldn't need the wig anymore. So it's kind of like in the middle.

15. **Teacher:**   That's a great point.

16. **Holly:**   I was thinking she could get a clear kind of string or something and she could tie it to her head.

17. **Teacher:**   Those are some really good ideas about how, if you chose the wig as the solution, you'd help solve some of the issues that could arise, such as it falling off, depending on her activities. That's really neat. So, for you, you might be leaning more toward this solution because you're finding ways to deal with some of the cons you're coming across. Very interesting. Did anyone else have a pro or a con for the wig solution? Lily?

18. **Lily:**   I think one of the cons ... If someone, when her hairs grows back all the way, and she could only find a red-haired wig, and if she had black hair, when her hair grows back, people would say, "Did you color your hair or something?"

19. **Teacher:**   Okay, so the wig might not match her hair.

20. **Dash:**   That's actually a good possibility.

21. **Teacher:**   Okay. Atticus?

22. **Atticus:**   Um, a con about the glue. If the hair grows back under it, it would really hurt to take the wig off.

23. **Teacher:**   Okay, so it could hurt to take it off.

24. **Jack:**   And the hair might get stuck on the glue, then it will yank the hair off right away.

25. **Holly:**   And it might just rip off before it even grows back.

26. **Teacher:**   So that would hurt, absolutely. Any more pros or cons about the wig? Claire?

27.	**Claire:**	I have a con. Sometimes wigs are really itchy. Wigs are usually really itchy, so ...
28.	**Teacher:**	I was thinking that, too. Wigs are really itchy. I don't know if you've ever worn one for a costume or something, and she doesn't even have hair to protect her, so it'd be itchy.

After fully discussing the wig solution, the class had similar conversations about a few of the other possible solutions they brainstormed. The teacher had them build a solution as part of the unit, but it was around a different problem. She used the wig problem as a place to support students as they practiced brainstorming and scoping the problem. A conversation such as this also helps students think about the constraints associated with the book and the classroom.

These introductory activities give insight into the engineering design process and help students build skills specific to engineering, such as iteration and testing. They also help set the stage for open-ended projects and a collaborative classroom culture. These skills will be important as students begin participating in engineering design activities that more closely represent the types of problems professional engineers deal with daily.

First Book Experience

We suggest picking a short picture book to be part of students' first Novel Engineering experience. During their first experience, students will try to determine your expectations and how to navigate the open-endedness of the task. If this is the first time you are doing Novel Engineering, it will also be a time for you to see what works and what does not. During this first task, not all students will understand that you want them to design and build a functional prototype. For some students, it is not until the end of the activity when they look around and see that many of the other solutions are functional that they realize what it means to make something work.

It is for this reason that we suggest starting with a short book so you don't invest a lot of time reading something longer, but you and your students can still gain insight into the experience. Good "first-time" books include *Muncha! Muncha! Muncha!* by Candace Fleming, *Peter's Chair* by Ezra Jack Keats, and *The Relatives Came* by Cynthia Rylant. We've found that even older students have fun with these books. Of course, if there is another short book that contains problems you and your students enjoy, go ahead and start with something you already know.

Classroom Culture

Novel Engineering will work well in a classroom that encourages creativity and curiosity but also has a clear set of expectations. In a supportive engineering design culture, students are free to tap into their imaginations, test out new ideas, and recognize failure as an opportunity to learn. Students should feel safe in making mistakes and taking chances. Students and teachers should be in constant communication, and critique should be embedded throughout the process.

Sometimes, the open-endedness of this type of project—coupled with the physical and intellectual messiness of engineering—is difficult for teachers and students to navigate; it is very different from what typically happens in a classroom. This open-endedness is not meant to be a free-for-all, so being clear about what everyone will be doing during a unit helps keep things under control. It's best to be explicit in your expectations when speaking with students, since this will help them understand and learn from the experience. Student expectations during Novel Engineering projects include the following:

1. Build a functional prototype.
2. Collaborate with a partner.
3. Understand and address the needs of characters/clients.
4. Provide constructive criticism.
5. Make changes based on test results and peer feedback.

Collaboration, Communication, and Critique

Collaboration, communication, and critique are three elements that are visible in a productive Novel Engineering classroom. These behaviors help students fully engage in a Novel Engineering unit and build skills that are beneficial in other academic and personal endeavors.

Collaboration builds on the perspectives, knowledge, and capabilities of team members to address design challenges. Since students are working in pairs or small teams, they are collaborating on all phases of the process. They are not only collaborating with partners but also members of other groups and the whole class. And just as with other new skills, students need to be taught to collaborate.

Collaboration includes the skills of communication, problem resolution, and task management. Many times, students assume they are collaborating by divvying up tasks and completing them independently. However, doing so may allow students to avoid communication or to grapple together in finding a resolution. We don't advocate for divvying up tasks, but it's important to note that some students (e.g., those with different learning needs or executive functioning deficits) may need tasks to be divided and assigned. Our hope is that, with multiple Novel Engineering experiences, these students will become better at identifying tasks and collaborating to complete them. Since students have various abilities to collaborate, groups may need a range of teacher support.

Effective communication allows students to collaborate productively and collectively understand the needs of a customer as they justify design decisions. Although students do not communicate directly with clients in Novel Engineering, they do communicate with one another throughout the design process. In Novel Engineering, students communicate orally, as well as through their written work and manipulation of materials.

Although communication and teamwork are often a focus for students' overall development, communication is not limited to getting an idea across. Students need to listen to and understand one another's ideas—not just "get along." For example, the students reading *From the Mixed-Up Files of Mrs. Basil E. Frankweiler* (Chapter 4) questioned one another, made decisions based on the quality of their ideas, and helped one another deal with setbacks. These actions helped them work as teams and ultimately improve their solutions.

The following student discussion shows that these skills are critical for groups to engineer a functional, usable device. You will again meet Mark and Charles (from Chapter 3) as they work on their solution for *James and the Giant Peach* and deal with a conflict regarding their ideas. If you recall, Mark and Charles used drawings not only to communicate their ideas to each other but also to make sure they understood each other's ideas. Although the drawing furthered their understanding, you'll see in this excerpt that they needed another way to further explain their design ideas. They are trying to figure out a way to raise the giant peach, but are having trouble figuring out a way for the peach to go as high in the air as they would like.

1. **Charles:** How about we tie something to something, and then put something heavier on it, so it will go down, and the other—the peach—would go up.

2. **Mark:** So, like, um—

3. **Charles:** Like a type of seesaw!

4. **Mark:** So you're thinking of this idea. Like a platform thing, and it can be longer. And so, you want the peach like that [on one end of the platform], and a net. And then you want an enormous rock right here to like lift that [on the opposite end of the platform].

5. **Charles:** No, actually, I want it like this. Watch. [*adds to his drawing*] I want a string tied to the peach, and then something heavier than the peach, lifting up.

6. **Mark:** That's kind of what I did right here.

7. **Charles:** You got what I'm saying all wrong! Okay, Look at this. [*grabs a water bottle and pencil*] Here's the water bottle, here's the peach, and here's the lever. Now, I'm thinking—instead of pushing the lever down, we put something heavy *on* the lever, so it will push the lever down.

8. **Mark:** But then it will have to balance, and then this will go up, and then the thing will fall off, and then this will go back down. That won't work. They need the peach to go all the way up. Yeah, so ... lever, peach. If they can make it like this.

9. **Charles:** Wait, here's the peach. Watch this. This would work. Watch. Pretend the peach on the other side balances. Now watch. Okay, you're right.

10. **Mark:** See, it falls and doesn't do it correctly. So how will that work?

11. **Charles:** It won't.

12. **Mark:** Yeah.

13. **Charles:** Then how about we use ... well, what are we going to do?

14. **Mark:** Um. Okay, your idea will definitely work. Unless the string is too skinny and they both snap. Thick. We need thick string, like rope. Thick rope.

Mark and Charles resorted to using manipulatives to better explain what they were envisioning. The boys attempted to explain their ideas by comparing them to real-world things such as a seesaw and a platform. This helped, but there was still confusion about what each of them was saying in relation to their design. Charles then grabbed a water bottle and pencil to show his idea and get Mark's input. By using tangible objects, the boys were able to communicate their ideas, which helped them understand each other and test their ideas. It also helped them understand how levers work and how to incorporate them into their design.

Communicating ideas and giving feedback are important parts of engineering, from elementary school all the way through professional engineering teams. Fostering students' ability to critique other students' work involves coaching them to offer helpful criticism that is both kind and specific. Just like collaboration, critique is a skill that needs to be taught and practiced. The classroom culture should therefore encourage students to give specific feedback that is both respectful and kind. Feedback can be positive, addressing the hard work students have done and celebrating their ideas, but it can also include questions and concerns specific to the design and context. Students should understand that the purpose of critique is to give information that will help others make constructive changes to their designs. With this in mind, the teacher should model thoughtful and specific critique when talking to students about their designs.

Critique should also provide students with information about what is working and not working in their designs. For example, instead of saying, "It looks good," you can say, "Your periscope looks like a lunch box, which helps the characters blend into the crowd." When offering suggestions, your critique should focus on one design feature at a time. For example, "The netting on top of your turtle protector looks like the spaces are too big, so the turtle will get out" is more helpful than simply saying, "The top doesn't look right."

The Teacher's Role

During Novel Engineering, the teacher should act as a facilitator, helping students realize their ideas and stretch as learners. During whole-group discussions, the teacher should provide suggestions and model collaboration and targeted skills. As students work, the teacher should walk around the classroom, talk to students, and ask questions about their designs. Teachers should spend time listening to, interpreting, and responding to their students' ideas (Hammer, Goldberg, and Fargason 2012).

A teacher's goal is to ask questions and offer comments that move students along (but not along a predetermined path) without solving the problem for them. We know there is a delicate balance between maintaining student agency and pushing them to expand their thinking and improve their designs. When you first engage in Novel Engineering, realize that you, too, are learning a new skill, so feel free to play around with phrasing and strategies.

Part of your work as a teacher is to let students know you are listening to their ideas and not looking for a specific or "correct" answer. To this end, a few questions and prompts we've used are as follows:

- Can you show me how this works?
- How will you be able to show the class that it works?
- Can you think of a new way to do that?
- Tell me more about it.
- What do you think you need to do to make this work?

- Why do you think it didn't work?
- So, you're saying that ...

In fact, it's okay to let students know if you don't know the answer to their question. Take the opportunity to learn together and model how to figure things out.

- I'm not sure if that would work. Let's try it out to see what happens.
- Hmm. I don't really know. Let's figure this out together.

Remember that the purpose of asking questions is to understand students' ideas and help them stretch their thinking. Therefore, you aren't looking for yes or no answers but asking open-ended questions they'll need to answer with a longer answer. If students are used to questions that have a specific right answer, they may need time to become accustomed to this form of questioning. Once they understand and are comfortable with the process, they will be eager to share their ideas with you and their classmates.

Assessment

Talking about assessment in a chapter about classroom culture may seem a curious choice, but the way students are assessed influences how they think about doing their work and what they think are acceptable solutions. If students' grades are solely based on the completion of a perfectly working solution, then they may play it safe and only want develop solutions that will work immediately. It is better to focus your assessment on the entire process and the conversations students have. Let them know that you value innovation and want them to take chances as they design.

Safety Notes

1. Wear safety goggles or glasses with side shields during the setup, hands-on, and takedown segments of the activity.
2. Use caution when using hand tools that can cut or puncture skin.
3. Use caution when working with heat sources (e.g., lamp) that can cause skin burns or electric shock.
4. Use only GFI-protected circuits when using electrical equipment, and keep away from water sources to prevent shock.
5. Secure loose clothing, remove loose jewelry, wear closed-toe shoes, and tie back long hair.

6. Make sure the trajectory for projectiles are marked off and do not allow anyone to stand in their path.

7. Immediately clean up any liquid spilled on the floor so it does not become a slip/fall hazard.

8. Wash your hands with soap and water immediately after completing this activity.

Reference

Hammer, D., F. Goldberg, and S. Fargason. 2012. Responsive teaching and the beginnings of energy in a third grade classroom. *Review of Science, Mathematics, and ICT Education* 6 (1): 51–72.

Website

Leveraged Freedom Chair: *http://gogrit.org/lfc*

Book Resources

From the Mixed-Up Files of Mrs. Basil E. Frankweiler; Konigsburg, E. L.; Age Range: 8–12; Lexile Level: 700L

James and the Giant Peach; Dahl, R.; Age Range: 7–10; Lexile Level: 870L

Muncha! Muncha! Muncha!; Fleming, C.; Age Range: 3–8; Lexile Level: AD560L

Peter's Chair; Keats, E. J.; Age Range: 3–7; Lexile Level: 500L

The Relatives Came; Rylant, C.; Age Range: 4–8; Lexile Level: AD940L

Planning a Novel Engineering Unit: Books and Materials

Chapter 10

In this chapter, we begin walking you through the planning stages of a Novel Engineering unit. We cover choosing a book, anticipating what students will do and talk about, and selecting appropriate supporting and building materials. A detailed document to help you plan a Novel Engineering unit is included as Appendix F (p. 231). If you want to see something a bit more specific, a sample unit guide for the book *America's Champion Swimmer: Gertrude Ederle* by David A. Adler (featured in Chapter 5) is available on this book's Extras page at *www.nsta.org/novelengineering*. We do not expect you to follow this guide step by step, but we designed it to be flexible and to give an idea of the structure of a Novel Engineering unit.

Choosing a Book

The open-endedness of a Novel Engineering unit allows teachers to choose books that are appropriate and interesting for their students. In Novel Engineering professional development workshops, we review books that educators are currently using and examine the advantages and challenges of using those books in a Novel Engineering unit. In this chapter, we discuss traits of books we've seen work successfully. Our aim is to help you anticipate what students will do in a Novel Engineering unit so you can prepare appropriate units and respond to students as they work.

Teachers have used books of different genres and lengths for Novel Engineering units, choosing books that meet the needs of their students and classroom goals. Identifying and choosing a book for Novel Engineering is similar to choosing one for English language arts (ELA). In short, the book should be

- age appropriate,
- interesting to students,
- complex enough to stimulate in-depth discussions, and
- inclusive of a variety of problems students can identify and solve with engineering.

Choose a book you've read before and know your students will enjoy reading. Books with characters who are relatable, either because of age or interests, allow students to identify with them, which helps them better address characters' needs and preferences as they design. For example, in the book *From the Mixed-Up Files of Mrs. Basil E. Frankweiler,* the characters Jamie and Claudia resonated with the fourth graders because they were of a similar age and had many similar feelings. In addition, the more complex the characters, the greater the chance of in-depth discussions and debate related to design choices.

As we said in the previous chapter, the relative simplicity of many picture books offers a great starting experience for students. For example, in *Muncha! Muncha! Muncha!,* students were able to infer characters' feelings and needs and to consider different perspectives, even though the book does not directly provide a lot of information about the characters' emotions or behaviors.

Genres that work well include realistic fiction, historical fiction, and nonfiction (particularly biographies). In Chapter 2, we saw a middle school class use a realistic fiction book, *A Long Walk to Water,* as they engineered solutions to help a young girl. When conducting our research, we saw another class use a nonfiction text about enslaved people in ancient Egypt. As students learned how the pyramids were built, they noticed the problem that many people were badly hurt or killed and were forced to endure great hardships during process. Students then did a Novel Engineering project to design and build something that would increase safety for the people building the pyramids.

We've found that books with magical elements or that are highly fantastical do not fit well for Novel Engineering units. Though they are fun to read, these stories often take place in worlds where solutions can be magical and the characters do not necessarily have to obey the real-world laws of physics. When using these books, students may design solutions that magically solve the characters' problems (e.g., a fortune-telling ball, a time-traveling robot).

Our aim with Novel Engineering is to give students opportunities to design both functional and feasible solutions—in other words, those that are testable in a classroom environment and make sense in the story. Although fantastic solutions give students opportunities to be creative, they make it difficult for students to develop engineering skills and follow the engineering design process. That said, we have seen students successfully engineer functional solutions for some fantasy books, such as the lever system Charles and Mark used as they designed for *James and the Giant Peach* by Roald Dahl.

Other books that do not work well include those without well-developed characters and interesting settings. An example would be a picture book about pollution and the effects of trash and factories on people. The book may be age appropriate, but if it lacks a character with whom students can connect, there's not much for them to consider as they prototype and test solutions.

Questions to Consider

Before your students read a book, we recommend looking at it through the lens of an engineer. Ask yourself, "Does this text include problems for which it is possible to engineer or build a solution?" For example, one problem that occurs in Roald Dahl's *Danny the Champion of the World* involves Danny's father getting stuck in a ditch. When considering this book, a teacher (or engineer) might imagine a number of engineering solutions to free Danny's father.

In contrast, a book in which most problems involve a character being bullied may not include many problems that are easily solved through engineering. Bullying is a social-emotional issue, which requires conversations about empathy and treating others with kindness. Engineering *may* help these characters resolve their problems, but we do not want students to walk away with the idea that an engineering solution will help them escape similar social dilemmas.

Books that have both engineering and social-emotional problems provide a fuller context for Novel Engineering units. Class discussions will help students determine if the problem they chose can be solved through engineering. In general, if you are not able to identify problems that can be solved by engineering, then the book is probably not a good choice for Novel Engineering.

Keep in mind, though, that the problems don't have to be central to the plot of the story and that problems can come in unexpected places. In *From the Mixed-Up Files of Mrs. Basil E. Frankweiler*, students identified the characters' challenge in seeing a statue over a crowd or getting additional money from a fountain. A solution to these problems may not change the course of the book, but they can be addressed with engineering and require students to engage with the text and understand the characters' needs.

As you think about the problems you have found in your chosen book, consider if they are age-appropriate problems for students to solve. Let's return again to the students we saw in Chapter 2 who read *A Long Walk to Water* and their solutions related to Nya's daily trips to get water for her family. Two groups chose to tackle the difficulty of hauling a sufficient amount of water over a long distance and walking on the thorny, rocky ground. Both problems involve relatable experiences that may connect to students' own lives (i.e., carrying heavy loads and stepping on sharp objects). In addition, students could solve both problems with available materials, such as rope, foam, cups, and buckets. By contrast, students likely would not have personal experiences to help them relate to a larger problem in *A Long Walk to Water*: the lack of water in Sudan. The sheer complexity and scale of this problem requires teams of engineers and other experts; it is not something students would be able to solve in a few class periods.

In Table 10.1, we've listed some of the books we've seen implemented in Novel Engineering units. Hopefully, this will inspire you to think about books to use and anticipate some of the problems and solutions your students identify. All these books offer problems that can be solved by students.

Special Considerations

The plots of some books require a deeper consideration of how Novel Engineering will be implemented. In a fourth-grade classroom, for example, students were doing a Novel Engineering unit with Kate DiCamillo's *The Miraculous Journey of Edward Tulane*. This book is about Edward, a porcelain rabbit who gets lost at sea. Several students wanted to design solutions to rescue Edward, but they encountered an unanticipated challenge when it came to testing their designs: the ocean setting is not something that can be easily replicated or modeled in a classroom.

The teacher was able to obtain a large, plastic container, which she filled with water to represent the ocean. However, the students voiced frustration as they began testing. Some groups used the sides of the container to make it easier to retrieve Edward (represented by a small, heavy toy) from the bottom of the container. Students quickly realized this would be impossible in the ocean because it does not have walls. The class then had to generate a revised set of constraints specifically related to testing, so they would at least all be testing under the same conditions.

Characters are another special consideration to keep in mind. Students typically analyze characters' abilities as they design solutions. If characters possess magical powers or qualities, then it may be difficult for students to think about solutions that do *not* call on the characters' magic. Students also love designing for animal characters. However, designing for animals can be tricky when there is uncertainty about what physical and mental capabilities the animals possess.

Table 10.1: Books used by teachers for Novel Engineering units

Book	Grade	Plot	Problems Identified by Students	Solutions Designed by Students
The Snowy Day written and illustrated by Ezra Jack Keats	1	Peter plays in a city snowfall and makes a snowball to keep. It melts in his pocket, which makes him sad.	• Peter wants to keep the snowball, but it melts in his pocket. • His snowball melts in the pocket of his snowsuit, so his snowsuit is now wet.	• Insulated snowball savers • Portable insulated snowball saver
Clementine written by Sara Pennypacker and illustrated by Marla Frazee	2	Clementine and her family live in an apartment building where her father is the building's manager. He is trying to keep everything clean and discourage pigeons from messing on the building.	Clementine helps her father keep pigeons off the building so they don't leave a mess.	• Owl with movable wings and attached megaphone that projects predator sounds • Fan with large blades that will deter pigeons from landing on the ledges and windowsills
The Relatives Came written by Cynthia Rylant and illustrated by Stephen Gammell	1–2	A family's relatives come from Virginia to visit for a few weeks.	There are not enough beds for the relatives to sleep on.	Tall bunk beds with ladders that will allow the entire family to sleep comfortably
The Three Little Javelinas written by Susan Lowell and illustrated by Jim Harris	1–2	Three little javelinas try to keep a coyote from eating them.	• The coyote is a predator and needs to get his meal. • The javelinas need to be protected from the coyote.	• House with trick chimneys that will discourage the coyote • House with a roof that has spikes on it so the coyote can't get to the chimney
If You Lived in Colonial Times written by Ann McGovern and illustrated by June Otani	3	This is a nonfiction book describing the lives of people in the New England colonies before the Revolutionary War.	• Transporting water and watering gardens is a lot of work. • Picking vegetables is difficult. • Colonists don't want to drink dirty water.	• Garden watering system using a central place to dump buckets of water and tubes for distribution • Hinged vegetable picker with a long handle • Filtration system for removing dirt from water

(continued)

Table 10.1: Books used by teachers for Novel Engineering units *(continued)*

Book	Grade	Plot	Problems Identified by Students	Solutions Designed by Students
The Invention of Hugo Cabret written and illustrated by Brian Selznick	3	Hugo is an orphan who works winding clocks in a train station. He uncovers mysteries as he tries to fix the automaton, a mechanical man, that he and his father used to work on.	• Hugo has to wind all the clocks on his own. • People are after Hugo because he keeps stealing food and other things.	• Clock winder that only requires Hugo to roll a ball down a tube • Booby trap to catch people chasing him
The Mouse and the Motorcycle written by Beverly Cleary	3	A mouse named Ralph who lives in a hotel takes his friend's toy motorcycle without permission and rides into a wastebasket by accident.	The mouse and the motorcycle need to get out of the wastebasket before the maids empty it into the trash incinerator.	• Ramp to drive the motorcycle out of the wastebasket • Rubber-band launching device to get the motorcycle out of the wastebasket • Pulley to lift the motorcycle out of the wastebasket
Pop's Bridge written by Eve Bunting and illustrated by C. F. Payne	3	Two boys watch their fathers build the Golden Gate Bridge in the 1930s and share in their fathers' pride and sense of accomplishment, but they also become aware of the great dangers they face while working on the bridge.	People fall off the bridge during its construction.	• Backpack that becomes a bungee cord • Parachute with parts that span the bridge
From the Mixed-Up Files of Mrs. Basil E. Frankweiler written by E. L. Konigsburg	4	Claudia and Jamie run away from home and stay at the Metropolitan Museum of Art in New York City. They face challenges such as how to wisely spend their money, avoid detection, and learn about a mysterious sculpture that may have been created by Michelangelo.	• Carrying around $24 in loose change is difficult and noisy. • They need to get around the city without paying too much. • They can only see the statue when it is surrounded by crowds.	• Backpack with a padded false bottom to hide money and muffle sound • Scooter made from found materials with a wheel system created from rows of ping pong balls • Telescoping periscope with adjustable mirror flaps

(continued)

Table 10.1: Books used by teachers for Novel Engineering units *(continued)*

Book	Grade	Plot	Problems Identified by Students	Solutions Designed by Students
The Gorilla Who Wanted to Grow Up written by Jill Tomlinson and illustrated by Paul Howard	4	Living in the mountains of Africa, Pongo the gorilla wants to grow up to be brave and strong like his dad. When his sister Whoopsie is born, however, Pongo realizes that growing up is not just about being robust.	• Pongo's sister hasn't yet learned how to climb trees. • Sleeping on the ground with predators hunting at night is dangerous. • Whoopsie gets stuck in a tree.	• Ladder that helps Whoopsie climb up the tree • Safe shelter that protects them from predators • Slide that attaches to a tree so Whoopsie can slide down
James and the Giant Peach written by Roald Dahl	4	James finds a magic peach that has grown to the size of a house in the garden of his aunts' house. With James and some large magic bugs inside, the peach rolls down to the ocean. When sharks take bites out of it, James figures out a way to fly the peach out of the water, but he then encounters new dangers.	• James is trapped by a tall fence at his aunts' house. • Cloud people throw ice balls at the peach as it floats by. • The peach gets stuck on the spire of a skyscraper.	• Prototype of a trampoline and figuring out how to position it so James does not hit the wall • Ice tube deflection system (a tube with a built-in slingshot to send the ice balls back at the cloud people) • Small-scale building block crane to lift the peach off the spire and place it on a truck
The Miraculous Journey of Edward Tulane written by Kate DiCamillo and illustrated by Bagram Ibatoulline	4	Edward Tulane, a china rabbit lost at sea, encounters many different people along the way back to his owner.	Edward falls into the ocean.	• Raft that Edward would fall into • Claws that would pick Edward out of the water
Number the Stars written by Lois Lowry	4	Ten-year-old Annemarie lives with her family in 1943 Nazi-occupied Copenhagen. She becomes involved in helping Danish Jews, including her best friend, hide and escape from the Nazis.	• The Danish Jews need a safe place to hide. • The Jews need resources in their hiding spot.	• Underground shelter that the Jews could climb into via a hanging ladder • Pipe delivery system to get limited amounts of water to the people in hiding

(continued)

Table 10.1: Books used by teachers for Novel Engineering units *(continued)*

Book	Grade	Plot	Problems Identified by Students	Solutions Designed by Students
Shiloh written by Phyllis Reynolds Naylor	4	Marty finds a beagle near his house but does not want to return him to his abusive owner, Judd. Marty tries to conceal and care for the dog, Shiloh, in the woods near his house.	Shiloh needs to stay in his pen so he stays safe.	• Protective shelter for Shiloh from which he couldn't escape
Tales of a Fourth Grade Nothing written by Judy Blume	4	Peter has a two-year-old brother, Fudge, who creates lots of mischief when he gets into Peter's room.	• Fudge goes into Peter's room and gets into his things, including Peter's pet turtle. • Fudge can get out of his crib and get into trouble.	• Cage that prevents Fudge from getting to Peter's pet • Alarm system attached to Fudge's crib that will ring a bell when he tries to escape
The Tarantula in My Purse: And 172 Other Wild Pets written by Jean Craighead George	4	A collection of nonfiction stories about a family that takes care of wild animals in their home.	• Some ducklings they take in must stay in the tub, so the family can't take a bath or shower. • When the ducks are put in the sink, they jump out.	• Mobile home for the ducklings that contains their food and water and has a cover so they can't jump out
The Trumpet of the Swan written by E. B. White	4	Sam helps a swan named Louis find a way to communicate, since Louis is mute. Sam also helps the swan in other ways.	• The eggs in the swan's nest face danger from predators. • It is difficult for Louis to fly with all his possessions.	• Shelter that covers all sides of the swan's nest with a door so Louis could get in and out • Glider that Louis would hold onto and fly with
Tuck Everlasting written by Natalie Babbitt	4	Winnie meets the Tuck family who cannot grow old because of water they drank from a spring in the woods. A stranger tries to get them sent to jail for kidnapping Winnie.	• If people find the spring, they might drink from it and become immortal like the Tucks. • Mae needs to escape from jail before she is executed.	• Water absorber that comes down from a tree and soaks up water from the spring if someone tries to go near it • Magnetic hook that would latch onto the jail key so Mae Tuck can unlock herself

(continued)

Table 10.1: Books used by teachers for Novel Engineering units *(continued)*

Book	Grade	Plot	Problems Identified by Students	Solutions Designed by Students
The City of Ember written by Jeanne DuPrau	5	Ember was built far underground centuries ago to protect a group of people from a war. The city's people do not know they are underground or that they need to return to the surface. Lina and Doon learn that Ember is running out of resources and figure out how to escape.	• Carrying Lina's toddler sister during the long trip out of Ember will be difficult. • There is no communication system between houses at night when the lights are off. • It will be difficult to transport everyone in Ember to the surface.	• Device to carry a toddler safely and comfortably • System to carry messages between houses using string, pulleys, and bells • Ferris-wheel-like system for carrying people up and out of Ember
Hatchet written by Gary Paulsen	5	After the small plane he is traveling in crashes in the Canadian wilderness, 13-year-old Brian has to survive on his own.	• Brian needs to stay warm in the wilderness. • Brian must find food.	• Shelter that traps in heat and prevents cold air from coming in with a bed that folds up using a pulley system • "Trap" tree that will catch an animal that walks by
The Most Dangerous Game written by Richard Connell	6	A professional hunter, Sanger Rainsford, is marooned on an island and becomes the prey of a fellow hunter, General Zaroff, who lives on the island.	• Rainsford is trapped by Zaroff in a tower. • Rainsford is hiding in a tree when the general comes and lights a cigarette directly underneath him, so he must hold his breath and not cough. • Rainsford could get stuck in quicksand.	• Retractable zip line that Rainsford can launch (sort of like a grappling hook) and then ride • Pipe to divert smoke from the main character hiding in the tree • Giant shoes to help Rainsford float on the quicksand swamp and not leave footprints

(continued)

Table 10.1: Books used by teachers for Novel Engineering units (*continued*)

Book	Grade	Plot	Problems Identified by Students	Solutions Designed by Students
Island of the Blue Dolphins written by Scott O'Dell	6–8	A young woman is stranded on an island after her tribe abandons their home. Karana has to survive, find shelter, find food, and protect herself from the elements and a pack of wild dogs.	• Karana must get food without using a weapon because in Karana's culture, women are not allowed to touch or use weapons. • Karana needs to protect herself from the elements and from wild dogs. • Karana needs to protect and store her food in order to keep it from other animals.	• Fish catcher with a long handle that has bait to lure fish inside and is designed to keep them inside • Functioning fence that Karana can use to protect her shelter from the wild animals • Pulley system so the main character can suspend food in baskets away from animals
A Long Walk to Water written by Linda Sue Park	6–8	This novel tells the stories of two 11-year-old children from Sudan. One story is about a boy who leaves 1990s Sudan to flee war. The other, set in 2008, focuses on a girl named Nya who walks through the desert twice a day to get water for her family.	• Nya has to walk over hot sand, rocks, and thorns. • Nya has to hold a heavy jug of water on her head. • The water that Nya and her community drink makes people sick.	• *Shoes that help protect Nya from the discomforts of the desert • Wagon that rolls to hold the water • Water filter to remove dirt and bacteria
Muncha! Muncha! Muncha! written by Candace Fleming and illustrated by G. Brian Karas	1–7	This picture book tells the story of the numerous ways a farmer tries to keep rabbits from entering his garden and eating his vegetables.	• The farmer wants to keep the rabbits out of his garden. • The rabbits want to get into the garden.	• Wall that surrounds the garden with a trap at the top, so if rabbits try to climb up and get in, the trap will fall on them • Climbing structure that allows the rabbits to hop on platforms until they reach the top of the wall surrounding the garden
*Weslandia** written by Paul Fleischman and illustrated by Kevin Hawkes	1–7	A boy named Wesley is bullied by his peers, so he decides to create his own civilization in his backyard complete with a staple crop.	Watering all his crops is a lot of work for Wesley.	Irrigation system that catches rain and releases water through straws and onto his crops

For example, in realistic stories, animals' abilities are straightforward; students either know or can look up the behavior of dogs, rabbits, or bears, for example. In other books, animals' abilities sometimes resemble those of a human; they talk, move, and behave like a human would.

In a third-grade classroom, students read *The Mouse and the Motorcycle* by Beverly Cleary. In this book, the main character, Ralph, is a thinking, talking mouse who drives a tiny toy motorcycle. As students read the book, they noticed the problem of Ralph falling into and getting stuck in a garbage can. A few groups of students decided to solve this problem by helping Ralph get out of the garbage can. Some used simple machines to help him out, such as pulleys, levers, and ramps, and these types of designs sparked conversations regarding Ralph's ability to use the devices. After listening to the students debate Ralph's abilities, the teacher helped the class come up with a list of assumptions regarding Ralph's abilities so they would all work and test under the same conditions.

Books that have anthropomorphized animals are not necessarily bad choices for Novel Engineering, but you should be aware that you will need to have some additional conversations with students to help them negotiate the characters' abilities. This can be an opportunity for students to return to the text to look for evidence that supports what they think the animals can do. In the case of Ralph the mouse, the students decided that he would be able to use his hands in the same manner that humans do. Textual evidence to support this assumption was that he drove a toy motorcycle and had to steer.

Choosing Materials

Anticipating the problems students choose to solve in Novel Engineering projects will help you consider and select appropriate materials. For example, if you plan to use *The Snowy Day* by Ezra Jack Keats, you might anticipate that many students will design solutions that will get wet. In this case, you should plan to have inexpensive, waterproof materials, such as plastic containers or cups, available in the classroom (see Figure 10.1, p. 174).

You can choose a wide variety of building materials for students to use in their designs. Novel Engineering was developed so projects could be done with materials ranging from inexpensive household items to high-tech materials and tools typically available in maker spaces (see Appendix A, p. 223, for a full list of suggested materials). Materials that are familiar, versatile, and easy to use will allow students to more confidently jump into designing. For example, corrugated cardboard is a great material for engineering projects. It is cheap, recyclable, and sturdy enough to hold its shape, yet it's malleable enough to be folded and cut into different configurations.

Figure 10.1: Building with Novel Engineering materials

The following is a list of materials we give to teachers at Novel Engineering professional development workshops. This list was developed by looking at the kinds of materials students most successfully used when building. It is a long list and intended to be a resource; we do not think a classroom needs to have all these materials. (A sample letter you can send to parents to request donations is included in Appendix G, p. 234.)

- Cardboard
- Tape
- Glue
- Plastic containers
- Cloth
- Aluminum foil
- Wire

- Fastening tape
- Modeling clay
- Pipe cleaners
- Dowel rods
- Rubber bands
- String
- Cotton balls

There may be tension related to including more technologically sophisticated materials. High-tech materials have a more significant "cool" factor, and their inclusion may allow students to design functional solutions, but there are often drawbacks related to cost and the time needed to learn the basics of the technology. In one case we observed, two groups from different classrooms approached the same problem using materials on opposite ends of the cost spectrum. The problem posed was how to help a dachshund with an injured spine who had lost the use of her hind legs. Both groups built a platform with wheels to allow the dog to walk. The first group used cardboard, tape, and premade rubber wheels. The second group used interlocking construction blocks with motors that were programmed to move the dog's legs (see Figure 10.2). Both were successful designs (as far as could be tested on a stuffed dog) that responded to the problem and the client's needs. Although the potential of the robotics kit led to a more sophisticated design, the kit costs several hundred dollars, and the time to build and test the solution took significantly longer.

Choosing which materials and tools to make available also depends on the setting and context of the book. You may choose as a class to limit the materials

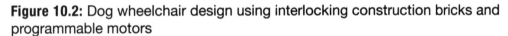

Figure 10.2: Dog wheelchair design using interlocking construction bricks and programmable motors

to what would be available to the characters. After students have identified problems and started brainstorming solutions, you should have a discussion with them regarding the materials they will use to build their designs. Open this discussion with a question such as, "What materials do you think would have been available to Jamie and Claudia in the museum?" Students then generate a list that might include cardboard, bottles, newspaper, rope, and string, and they might state their rationale for how the characters got those materials (e.g., "They found them in the recycling bin or trash"). Depending on your students and the text, you may also decide not to think too much about the materials characters would have access to and simply choose materials with which your students are most comfortable.

In Chapter 6, we describe what happened in a third-grade classroom with the book *If You Lived in Colonial Times* by Ann McGovern. Several groups of students assumed they were designing a solution as if they lived during the colonial period and discussed which materials could represent more historically accurate materials. Although it would be great if we had all the materials students could possibly need, it sometimes makes sense for them to use alternatives that have the same general properties. In that third-grade class, students did not have access to woolen material, so using cotton balls allowed them to pursue their design ideas with a material that has similar properties.

It is obviously beneficial for students to have had experience with the materials before they begin planning. Students who have limited experience with any kind of hands-on making may benefit from a chance to explore how craft and recyclable materials can be assembled in unique ways. In addition, if you feel that your students could benefit from a more complex understanding of the properties of the materials and how they interact with one another, refer to Chapter 9 for suggested activities.

When to Present Materials to Students

In addition to deciding which materials to use, you also have to consider *when* you will present the materials to students (see Table 10.2). There are trade-offs related to when you unveil the available materials. If students know exactly which materials will be available to them, brainstorming and planning can be tailored to incorporate the materials they see. The drawback to this is that students' ideas may be constrained based on what they think they are able to do with the materials. By contrast, if materials are presented to them after they have planned, their solutions may be more diverse and creative, but they may need additional time to alter their designs based on the available materials. In some cases, students might become stuck on their design and have a difficult time reimagining details.

Table 10.2: When to present the materials to students

	Potential Benefits	**Potential Negatives**
Before Planning	• Designs can be tailored to incorporate available materials. • Students do not become fixated on materials.	• Brainstorming may be limited because students are constrained by what they think they can do with the materials. • Students may need to rethink their chosen solution, which will take time and may not be possible depending on what is available. • Materials can drive design.
During Planning	Students have time to brainstorm ideas and make changes based on available materials.	• Design plans may be limited because the designs are constrained by what they think they can do with the materials. • Students may need to redesign based on available materials.
After Planning	• There is a greater chance that students' planned designs will be innovative. • Building may take more time because students might need to reconfigure designs based on what is available.	Students may struggle with or be unable to build their design with what is available.

Although we've seen teachers present materials at different times during the process, we feel that, in most cases, it makes sense to let students know what materials are available before or while they are planning. Waiting to unveil materials until after students have spent significant time planning can lead to the need for redesign based on available materials. This can cause frustration, the need for additional planning time, and in some cases a complete change of the proposed solution. A good way to balance these constraints is to have students brainstorm a few possible solutions for a short period of time (approximately 5–10 minutes) and then show them the materials so they can decide on a feasible solution with the given materials. If you use a similar set of materials each time students do Novel Engineering, their familiarity will allow them to better envision what types of solutions they can design. In cases when you will be including new materials, especially if you are planning to use robotics or some high-tech options, let the students know before they begin planning.

In some instances, materials can drive the design. This can happen at any point of the planning or building process. In one case, a pair of students wanted to use a robotics kit in their design. They chose a problem based on their desire to use the robotics, rather than because it was meaningful to the story's characters or themselves. In another classroom, a group's design called for a helium balloon. Once the class saw the balloon, several other groups decided that a balloon was integral to their designs and then changed their designs to incorporate helium balloons, regardless of whether the balloons would improve their designs or not.

In most cases, students are told they cannot take materials until they have drawn a sketch or filled out some sort of planning sheet. This ensures that students do not start grabbing materials with little thought to what they will design. Another method teachers have employed is to give each group of students the exact same materials. This is a more equitable way to deliver materials, but it can dampen students' creativity and the diversity of solutions. In most classrooms, we have noticed students are able to gather materials in ways that are fair and equitable. This occurs most smoothly when norms are in place for helping students decide how to use materials efficiently and consider their classmates' needs.

How to Present Materials to Students

There are a few ways to present materials to the students. The method you choose really depends on your classroom space and what you feel most comfortable with. The most common way is to spread out the materials on a table and provide time for students to walk around and look at everything. Alternatively, you could keep materials in plastic tubs and spread the tubs out on the floor or on a table. Another option is to make a list of materials and give each group of students a copy.

Kristen Wendell, a faculty member at Tufts University, has developed a mobile maker workshop that allows for storage and display of materials (see Figure 10.3 and *https://communityengineering.org*). The materials are kept in clear hanging bags that are organized and labeled by type. The bags are kept on a rolling coat rack that can be moved around the room or to different classrooms.

In this chapter, we walked you through things you need to think about as you plan a Novel Engineering unit. Although we recommend that you have a good idea of what you and your students will be doing throughout the unit, we hope that you keep things flexible and let the process be fluid. In the next chapter, we

Figure 10.3: Portable maker workshop

continue the conversation about planning, and focus on anticipating what students will design and how to facilitate the design process.

Safety Notes

1. Wear safety goggles or glasses with side shields during the setup, hands-on, and takedown segments of the activity.
2. Use caution when using hand tools that can cut or puncture skin.
3. Use only GFI-protected circuits when using electrical equipment, and keep away from water sources to prevent shock.
4. Secure loose clothing, remove loose jewelry, wear closed-toe shoes, and tie back long hair.
5. Immediately clean up any liquid spilled on the floor so it does not become a slip/fall hazard.
6. Wash your hands with soap and water immediately after completing this activity.

Book Resources

A Long Walk to Water; Park, L. S.; Age Range: 10–12; Lexile Level: 720L

Danny the Champion of the World; Dahl, R.; Age Range: 8–12; Lexile Level: 770L

From the Mixed-Up Files of Mrs. Basil E. Frankweiler; Konigsburg, E. L.; Age Range: 8–12; Lexile Level: 700L

If You Lived in Colonial Times; McGovern, A.; Age Range: 7–10; Lexile Level: 590L

James and the Giant Peach; Dahl, R.; Age Range: 7–10; Lexile Level: 870L

The Miraculous Journey of Edward Tulane; DiCamillo, K.; Age Range: 8–11; Lexile Level: 700L

The Mouse and the Motorcycle; Cleary, B.; Age Range: 7–10; Lexile Level: 860L

Muncha! Muncha! Muncha!; Fleming, C.; Age Range: 3–8; Lexile Level: AD560L

The Snowy Day; Keats, E. J.; Age Range: 4–8; Lexile Level: AD460L

Planning a Novel Engineering Unit: Facilitating Design

This chapter discusses how to structure the introduction of Novel Engineering problems and student work, how to anticipate students' ideas, and how to interact with those ideas so students both meet academic goals and follow their own ideas. The Novel Engineering Unit Planning Document included in Appendix F (p. 231) will help you think about facilitating design in your classroom.

As you plan, keep the Novel Engineering trajectory in mind (Figure 11.1, p. 182), using it as a guide rather than a step-by-step checklist. After students have identified problems, they need to scope those problems and design solutions. Partners should pick a problem from the list generated by the class. They then begin balancing the complex tasks of defining the problem, brainstorming possible solutions, and planning their chosen solution. These steps of the planning process are fluid, and students may move from one to another and then back. This is natural and aligns with how professional engineers work.

When you work with students, we suggest picking one problem from the list and conceptually planning a solution as a class so you can model and foster what you want students to do. For ease of discussion, we break the steps into distinct chunks as we discuss them. In Chapter 12, we go into detail regarding the method of reading and keeping track of problems. For this chapter, though, we look at brainstorming as part of scoping problems and designing solutions.

Figure 11.1: Novel Engineering trajectory

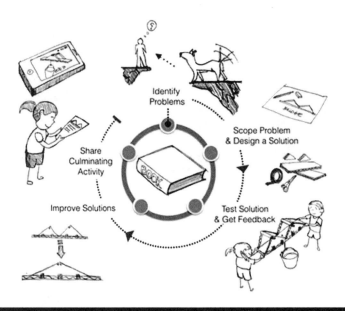

Grouping Students

As with all group activities, the dynamics of student partnerships play an integral role in a group's success. Novel Engineering requires a high degree of collaboration, and we encourage you to use your usual strategies for pairing students. Nevertheless, we will provide additional information to consider for design-related activities.

We suggest grouping students into pairs for several reasons. If students are not used to open-ended problems, it is often difficult for them to collaborate in a productive way with several other students. When students work in pairs, they are more able to manage tasks and negotiate ideas. Furthermore, in pairs, both students are able to physically work on the project at the same time—that is, they can both have their hands on the project. With larger groups, there is a possibility that some students may feel left out or end up watching others build.

Though it may seem like larger groups of students are easier to manage, teachers experienced with Novel Engineering have found that smaller groups function more effectively and make it easier for them to attend to individual interactions, whereas larger groups often need more help with personal interactions. When it is not feasible to have students work in pairs, we encourage you to spend time establishing norms for group work to help students remember to share and listen to one another.

There are many options for pairing students. To figure out which students to pair, we defer to your knowledge of your students and how they interact with one another. If having students choose their own partners or work with their tablemates works well for other projects, it may also work well for Novel Engineering projects. Another option is pairing students based on their interest in specific problems. For example, once the class has generated a list of problems from the text, ask students to list three problems they'd like to solve in order of preference. You can then pair students based on their desire to solve the same problem. In the case study about Gertrude Ederle (Chapter 5), the teacher shared her thoughts about pairing students based on how talkative they were. This is yet another way to form groups.

We are aware that it can be difficult for some students to work with others for a variety of reasons. Some students may need more explicit instructions about the engineering design process (EDP) and what they should be doing. We do not suggest making the EDP a checklist, but students who need executive functioning or social support may benefit from having a written list or diagram that names specific tasks and helps them recognize when they have sufficiently completed each task. It may also be beneficial to explain the process for coming to a consensus about design ideas. You may need to help partners negotiate who will work on specific tasks, such as collecting materials or taping together pieces of cardboard. Remember that even though students may work independently on separate parts of the design, they must still have conversations about how everything fits together. It is therefore important to have a discussion before the building work begins to avoid misunderstandings and confusion later. Hopefully, some of this proactive planning will help all students participate to the best of their ability.

Discussing and Categorizing Problems

As students read, have them keep track of problems in the text. (See Table 11.1, p. 184, for sample problems from specific books.) Then, during whole-class discussions, keep a running record of the problems students identify. Be prepared to hear a variety of problems—only some of which will be appropriate for Novel Engineering projects. Some problems can be solved with engineering, whereas others might strike you as "non-engineering" and more of a social-emotional nature. This is completely appropriate at this point because conversations that are social and emotional will lead to a better understanding of the book and help students arrive at well-thought-out design criteria and constraints.

Table 11.1: Sample books and student-identified problems

Book	Problems Identified by Students
Shiloh	• Shiloh is not safe in the pen that was built for him. • Marty has to do manual labor.
The Gorilla Who Wanted to Grow Up	• Whoopsie is too small to climb trees. • Sleeping on the ground at night is unsafe.
Hatchet	• Brian is cold sleeping in the wilderness. • Brian is hungry and needs to find food.

Older students should be able to differentiate between these two types of problems when asked to choose one to solve. Younger students will likely need additional support to figure out which problems are more appropriate for engineering solutions. One way we've seen teachers deal with this is to make two columns on the board with the headings "can be solved with engineering" and "cannot be solved with engineering." They then ask students to put the problems in the appropriate category. This is usually done as part of a whole-class discussion. These classroom discussions can serve as a way to assess which students are struggling and to figure out what kinds of problems they can solve in the classroom. Whole-class brainstorming often helps students better understand what you are asking them to do. You can structure the conversations to include examples of possible engineering solutions that could address the listed problems.

As the class categorizes the problems, ask them why they think problems belong in one category or the other. Even if you feel that a problem is more social-emotional than engineering, we encourage you to hear students' ideas rather than assume the problem can or cannot be solved with engineering. Many times, students brainstorm engineering solutions and show that a problem can be creatively solved with engineering. For example, recall the backpack students designed after reading *From the Mixed-Up Files of Mrs. Basil E. Frankweiler*; it served a double purpose—to hide change and to alleviate Jamie and Claudia's arguments about saving and spending money.

Even after students have a handle on which problems can be solved by engineering, they may need to further refine the list based on classroom constraints (see Figure 11.2). Think about Nya's problems in *A Long Walk to Water* (Chapter 2). Solving the water crisis in Sudan is not something students would

Figure 11.2: Narrowing problem choices

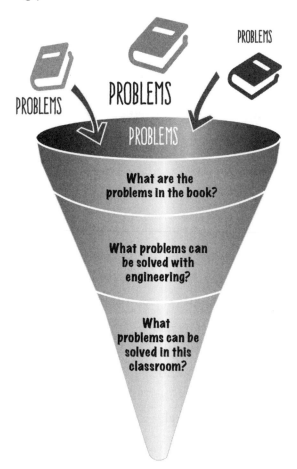

be able to do given classroom resources, but designing a pair of shoes is within the constraints of the classroom.

Choosing Problems

You will also need to decide if students will choose or be assigned engineering problems to solve. We recommend letting students choose their problems so they have more investment in and ownership of the process. There are three main options to consider: (1) all students solve the same problem, (2) students choose from a limited selection of problems, or (3) students choose a problem from the entire class-generated list.

With younger students or those who have not yet engaged in Novel Engineering, it may be beneficial to choose the first option and have all students solve the same problem. Some teachers feel this is a good way to first experience Novel Engineering. You can still support students' agency and let the class choose the problem they will solve from the list they created. This allows you to support all students more easily since whole-class discussions will relate to what each group is doing—as opposed to if they were all working on different problems. This approach also allows students to see that there are multiple solutions to the same problem.

By contrast, older students or those who have Novel Engineering experience should be able to choose the problem they want to solve. Allowing students to pick the problem helps them feel more ownership. In addition, some teachers find it easier to keep track of student progress and group conversations when they are working on different problems and solutions.

What counts as an engineering problem? It may be helpful to ask students a few questions to help them determine if the problem they chose is solvable with engineering:

- What problem are you solving?
- Who does that problem affect?
- What is your proposed solution?
- How will this solve the problem?
- How will we know it if works?

We have discussed a range of different engineering problems and solutions throughout this book, including a periscope for children to see over adults (Chapter 4), a device to help a long-distance swimmer eat during her swims (Chapter 5), and a hearing aid that is hidden in a backpack (Chapter 8). We classify these problems and solutions as "engineering" because the students designed creative, functional solutions to help characters from a book, and they could show that their solutions worked through testing and evaluating.

We have also seen many non-engineering solutions in classrooms. Even though these solutions are creative and may help a character, they often focus on aesthetics rather than function. For example, a student may want to embroider a blanket to make a character feel better or decorate a mask to help a character be inconspicuous. Although both of these solutions address a problem, they are not engineering solutions since they are cosmetic alterations and not functional prototypes capable of being tested and iterated.

No matter how much preparation teachers provide and regardless of how well they clarify expectations, some students will still offer solutions that are not engineering solutions. Nevertheless, that's part of the learning process. Try to understand students' ideas, see if you can figure out why they are interested in a particular problem or solution, and try to channel those interests into a similar—but functional—solution.

Brainstorming

Brainstorming, the process of generating ideas for a solution, is a valuable part of the EDP. It allows students to be creative and explore ideas before committing to them. Keep in mind, though, that there is not a one-size fits all approach. Deciding how to structure brainstorming in your classroom is guided by your knowledge of your students. Do your students need time to come up with ideas on their own before working in a group? You may want them each to sketch a few solutions and then share those with partners. Do you think your students might struggle to come up with functional ideas? They may benefit from whole-class collaborative brainstorming that generates multiple ideas from which groups can choose. In this section, we describe strategies for encouraging brainstorming throughout the EDP, helping students work collaboratively while brainstorming, helping students keep track of ideas, and setting up a short class discussion for students as a way to practice sharing ideas with one another.

When students are asked to brainstorm solutions to a problem, they typically write down several—maybe three or four—possible solutions. While students may list a variety of solutions, they may already have one solution in mind and just list additional ideas in order to complete the assignment. Forcing brainstorming at the beginning of the EDP is not always the most generative approach. Brainstorming is present throughout the entire EDP, and it can sometimes be more productive for students to generate ideas after they have begun designing. Even after students have chosen a solution, that solution my change as they consider constraints and criteria. We talk about brainstorming as a standalone step, but remember that it is a messier endeavor than just listing possible solutions.

To move students from being a novice engineer to having a more complex understanding of what they should do as they brainstorm, model brainstorming in your whole-class discussions before students start working in pairs or groups. Pick one problem from the class-generated list and have the class brainstorm several solutions. In the *Clementine* example (Chapter 3), we describe a second-grade class that brainstormed several options to one problem and then picked one possible solution—a wig—that was the focus of a discussion about pros and cons. As students named solutions, they also discussed the pros and cons associated with each. This whole-class exercise of brainstorming and identifying pros and cons should help students understand what they will be doing and give them experience to use when they begin working with a partner.

We've seen a few strategies that help students keep track of their ideas. Sticky notes allow students to record and manipulate ideas as they discuss feasibility with their partner. The notes are concrete representations of ideas that can be moved around, based on concepts such as identified criteria. This is helpful to students with organizational issues, and they help facilitate conversation between partners and develop a shared understanding of which ideas are the best possibilities. A worksheet or engineering journal provides a more structured approach to brainstorming. (See Appendix H, p. 235, for a brainstorming worksheet and Appendix I, p. 236, for additional examples of problem scoping, planning worksheets, and student work.) It is helpful if you allot time for students to share their design ideas before they begin building. This will allow you to hear each group's ideas and ensure feasibility. It also means students will be able to hear one another's designs and can practice providing critique and asking questions.

Planning

As with brainstorming, you can give your students their first taste of planning when they are involved in a whole-class discussion around one specific problem. Doing this first as a group will help students figure out what they need to think about as they plan—and what is expected of them. There are many ways for students to plan, and different methods allow students with different learning styles to find one that best facilitates communication and thinking for them. When thinking about how your students will plan, think about the following questions:

- Which mode will they use to plan?
- What information will they provide when planning?
- Will they plan individually or collaboratively?
- Whom are they planning for?
- When will they plan?

Which mode will they use to plan? Teachers often give students a planning document or have them work in an engineer's notebook to provide some structure. (See Appendix I, p. 236, for examples of planning worksheets.) These documents offer students a space to record their ideas through text and drawings. Usually, students talk to their partners as they plan, even when participating in another mode. It's important to note that while students are engaged in conversations, they may appear off track, but their thinking is tangential to their design plan. Remember that brainstorming often bleeds into planning, so this may be part of their process.

Other students might need to manipulate physical objects to think through how their ideas would play out in the real world. For example, when planning a design solution for *James and the Giant Peach*, Mark and Charles grabbed a water bottle and pencil to explore how a lever would work in their design. Likewise, some students may need to touch the provided materials to help them think through design ideas before they can commit to a specific design feature. These different modes are not mutually exclusive, so some students may engage in all of them.

What information will you ask them to provide when planning? Think about what information you want students to relay through their planning documents. If you use a worksheet, it should help them structure their thinking and communicate their ideas. Worksheets are an opportunity to communicate your expectations for what students should be doing. For kindergarten and first-grade students, worksheets should be brief and focus less on making inferences (see Figure 11.3, p. 190). Be conscious of the amount of time it takes students to fill out your worksheet; remember that they need sufficient time to construct, test, and revise their designs. The following are a few sample questions that may be useful in helping your students plan:

- What problem are you solving?
- Who is this a problem for?
- What is your solution?
- What are the pros and cons of your solution?
- Did you think about another solution? Why did you not choose it?
- What are the design constraints?
- What are the design criteria?
- What materials do you need?
- How will it work?
- How will you test it?

Will they plan individually or collaboratively? Since students are working together, they must collaborate and come to a decision about what and how

Figure 11.3: Planning worksheet for first graders

Names: _____

What could you build to help Peter keep his snowball longer?

Please write and draw about what you would make.

Source: Tufts University Center for Engineering Education Outreach.

they are building (see Table 11.2). For some partners, this may be a collaborative exercise; others may benefit from having individual time to think and then sharing and discussing their ideas with a partner. The goal is for students to arrive at a consensus about what they are going to build.

Whom are they planning for? Consider who the intended audience of the planning documentation will be. This will guide the format and content of the planning document. For example, will the document be used to facilitate conver-

Table 11.2: Collaborative versus individual planning

Planning Task	Possible Benefits	Possible Drawbacks
Working collaboratively on one document	• Students need to agree on their idea and what is recorded on the planning document. • Students need to communicate about what goes on the paper, which can trigger new and more robust ideas.	Students may not agree about what to write down, and they end up arguing.
Working independently on individual documents	Students do not need to spend time negotiating what should be written down.	Students may have completely different ideas, and extra time will be needed to reconcile their ideas.

sation between partners? Is it to show the teacher what students will be doing? Will students need to be specific about the quantity and types of material on the planning document? Knowing the answers to these questions in advance will help you create the most beneficial planning documents and help guide students as they plan their designs.

When will they plan? Generally, students choose a problem and plan a solution after they finish reading the book. However, in some cases with early elementary students, it is advantageous for students to begin a Novel Engineering activity before they actually finish reading. For example, if the character overcomes all the problems by the end of the book, it might not make sense for students to help that character solve a problem for which there is already a solution. In these cases, we recommend teachers stop reading after the problem has become apparent and have students begin the brainstorming and planning processes. This will often be the case with younger children's books and picture books in which there are only one or two problems—that tend to get resolved by the end.

Setting Constraints and Criteria

In addition to constraints in a book, there are numerous constraints inherent to classrooms. There may be limitations or restrictions that govern the amount of time, money, and support teachers can allocate to one project. These constraints are not limited to school classrooms, either; they are real-world challenges that professionals of all disciplines must account for. We believe students benefit from being aware of constraints and learning to work with them.

As students begin the design process, remind them of the time and resource constraints, and help them come up with strategies for working within those constraints. For instance, if there is a limited amount of tape or glue, students might set up a tape/glue station to prevent a single group from taking all the tape or glue. Some examples of constraint reminders include the following ideas:

- You must build a functional design based on the materials available to you in this classroom (unless you will allow students to bring materials from home or outside).
- You should test your design and make changes based on your tests.
- You must build one design together with your partner/group.

Students should also identify criteria for their designs. Criteria are standards against which their design should be compared to show that it works and meets the user's needs. Criteria are important because they help students make decisions regarding their design, including materials, testing, and evaluation. In Chapter 2, we saw how Amos Winter identified several criteria, including cost and weight, that influenced his design decisions. Just as with professional engineers, students need to outline the measures of success for their solutions.

You need to decide if you will require students to work with additional constraints or if students may decide on the constraints they feel are important. An example of additional constraints would be requiring students to use materials that would be present and available in the setting of the book. Many constraints and criteria are based on individual solutions. For example, a group building a water filter for *A Long Walk to Water* may decide that their filter should be able to filter water at a specific rate. This criterion for success would not make sense for a group designing shoes for Nya, the main character. That group may instead decide that their criterion for success is that the materials must be lightweight but sturdy enough to walk across rocks with little discomfort to the user.

Additional Considerations

Another factor that comes up as students plan is **scale**. Some solutions can be built to scale (e.g., shoes for Nya), whereas it is impossible for others because of time, space, or money. Try to anticipate what your students will propose as solutions and the appropriate classroom scale. Small-scale designs should still have a functional component that can be tested so they do not become representational models.

When students have a range of solutions, they also need diverse testing situations. Some tests will be obvious (e.g., Do the shoes you built stay on your feet?), but you may need to help students brainstorm appropriate testing situations when it is less obvious (e.g., How do you keep an animal in a pen without using real animals?). In these situations, help students figure out which design criteria they will test, along with an appropriate test.

Another situation you should anticipate is needing to think about testing and scale. In the second example just mentioned, students won't likely build a life-size animal pen. Even if you could bring a live animal into class, the actual structure would probably not be large (or sturdy) enough to test. Help students look at their list of design constraints and figure out which constraints can be tested at the scale at which they are working. In this example, the teacher found small, battery-powered animals for students to use as they tested the sturdiness of their smaller-scale pen.

If you feel unsure about how you can anticipate all these classroom considerations the first time you do Novel Engineering, remember that this is a learning process for you, too, and that you can narrow the target skills so that your students don't need to do everything the first time. For example, the first time, you could pick a book and focus on finding problems with your students. The next time, your students could pick problems and do a short planning activity. Even when you do the full building experience, it's fine for students to focus on some portions over others.

We've already talked about how the classroom culture can be structured to **support student discourse**. There will be multiple times throughout a Novel Engineering unit when students' conversations will organically offer opportunities to engage in meaningful discourse around engineering, literature, and other disciplines. As you plan, think about times in the process when you want to ensure that these conversations occur. For example, you may want to schedule time to discuss critiques or address testing with the class if those conversations have not yet occurred naturally. Overall, anticipating as much as possible what students might do will help you respond to what they end up doing and discussing.

In Chapter 9, we talk about **assessing student work** and how the way students' designs are assessed influences their participation in open-ended design work. While we were conducting research for this project, we spoke with teachers about how they wanted to assess the engineering portions of their projects. Many said that engineering is the one place in the curriculum where students can help guide the direction of their experience. They also said they were able to clearly see students' strengths and weaknesses when they were doing Novel Engineering, and this helped them support students in other ways.

We realize that many teachers must include some sort of assessment as part of a Novel Engineering unit. *Next Generation Science Standards* provide criteria that can be used as a benchmark for students. If you want to develop a form of assessment that helps you look at students' individual projects, you can build it around how closely their designs include certain critical aspects:

- Is the design functional?
- Does it function consistently?
- Does the design address the problem that was chosen?
- Does it work in a way that helps solve the problem?
- Is the solution appropriate to the book, its setting, and its characters?
- Were the materials appropriate to the task and the book?
- Does the design meet the criteria that were outlined when planning?
- Does the design consider the constraints that were outlined?
- Did students test their design?
- Were design changes made based on feedback from peers and testing?
- Were students able to communicate their ideas about the design to their partners and to the class?
- Did students copy an existing design or was it a completely new idea?

These questions can help reveal what students are thinking and building; the questions can also guide students toward what you want them to think about as they design and build. The range of questions also allows students to do well even if their final solution does not work perfectly. It shows them that you value the ideas behind their designs as well as the functionality.

Safety Notes

1. Wear safety goggles or glasses with side shields during the setup, hands-on, and takedown segments of the activity.
2. Use caution when using hand tools that can cut or puncture skin.
3. Use only GFI-protected circuits when using electrical equipment, and keep away from water sources to prevent shock.
4. Secure loose clothing, remove loose jewelry, wear closed-toe shoes, and tie back long hair.

5. Immediately clean up any liquid spilled on the floor so it does not become a slip/fall hazard.

6. Wash your hands with soap and water immediately after completing this activity.

Book Resources

America's Champion Swimmer: Gertrude Ederle; Adler, D. A.; Age Range: 4–7; Lexile Level: 800L

From the Mixed-Up Files of Mrs. Basil E. Frankweiler; Konigsburg, E. L.; Age Range: 8–12; Lexile Level: 700L

James and the Giant Peach; Dahl, R.; Age Range: 7–10; Lexile Level: 870L

If You Lived in Colonial Times; McGovern, A.; Age Range: 7–10; Lexile Level: 590L

Shiloh; Naylor, P. R.; Age Range: 9–12; Lexile Level: 890L

A Long Walk to Water; Park, L. S.; Age Range: 10–12; Lexile Level: 720L

Clementine; Pennypacker, S.; Age Range: 6–9; Lexile Level: 790L

Planning a Novel Engineering Unit: Literacy Connections

In this chapter, we focus on planning to support literacy connections in Novel Engineering. As stated before, there is not one "right" way to structure the types of literacy tasks you include in a Novel Engineering unit. Your choice should be centered on the literacy goals you have for your students. As outlined in Chapter 3, we consider literacy to include reading, writing, and speaking, and Novel Engineering can support all three modes of communication throughout the trajectory of a unit (see Figure 12.1, p. 198).

Literacy Connections

In previous chapters, we touched on how to facilitate discussions about problems and how to link them to specific characters. In this chapter, we don't discuss specific literacy programs or strategies but rather help you anticipate what you will need to prepare for a Novel Engineering unit. We touched on some of this in Chapter 11 since much of what students do covers both engineering design and literacy. For example, students may fill out a detailed planning document that helps them think about the mechanics of a design but also requires them to write about how the design will work.

Novel Engineering units include several opportunities for students to engage in literacy activities. As you've read, the Novel Engineering trajectory begins with students reading a book, finding problems, and discussing those problems and how they relate to the characters and plot. Students can keep track of the problems on a worksheet (see Appendix J, p. 242), and they can discuss how

Figure 12.1: Novel Engineering trajectory

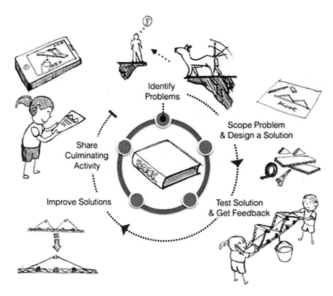

problems relate to individual characters (and make inferences so their designs address characters' needs and likes) with an empathy map. As they move to problem scoping, a planning worksheet helps them organize their thoughts as they brainstorm and begin planning their solution. Depending on the design, groups may use a worksheet to keep track of test data and outline desired changes based on the data. Students also have the opportunity to work on their oral communication skills when they present their designs during the mid-design share-out. As they complete their designs and begin wrapping up the unit, students can show evidence of their learning through a final presentation, an essay, or another type of culminating activity (e.g., a comic strip, an advertisement for their solution) that focuses on their reflections, the process they followed, or a mix of both.

Preparing for Reading and Discussions

Before your students interact with a book, read it on your own and mark stopping points in the text that highlight important concepts, vocabulary, or plot and character moments. This will help you think about the parts of the book you want to focus on in class discussions and anticipate students' questions and conversations that may spontaneously arise. You want to use these pauses as moments to make sure students understand the key themes and nuances of the book—as you normally would in a language arts lesson. In Novel Engineering, however, it can be useful to focus on students' understanding of the characters and settings so they can use that information both to understand the book and to inform their engineering. If you are using a picture book, it can be useful to pause on illustrations that help students infer information about the character or setting. Stopping points for longer books should be used to make sure students are tracking problems and raising any questions they have about the characters or setting.

There is no best way for reading groups to be structured in Novel Engineering. We've seen it set up a variety of ways. When students first engage in a Novel Engineering unit, we suggest reading the text aloud so you can model how to think about the relationship between the text and the engineering. This will help students understand what they will be asked to do, but after this initial exposure, it's appropriate for students to work in reading groups or read individually and then have small-group discussions. To allow for differentiation of students' reading levels, you can have groups read different books.

We suggest a combination of whole-class and small-group discussions. The former give you the chance to model conversations, collaboration, and critique. They allow you to guide discussions and make sure that crucial concepts and themes are heard by all students. In the latter, all students get the chance to talk and there is more back and forth during the discussion.

When students first do Novel Engineering and are engaged in small-group discussions, you will want to provide them with conversation starters to give an idea of what types of information they should cover. These prompts are not meant to guide the trajectory of their conversations but to act as a resource to facilitate discussions. You should model conversations with these prompts during whole-class discussions, but it's also helpful to write them down on the board or include them as part of a planning document. Examples of conversation starters include the following questions:

- What is the problem?
- Who was this a problem for?
- Why is this a problem?
- Does it affect other characters?
- What is a possible solution to this problem?

Planning for Writing and Documentation

Writing tasks can certainly be multiparagraph expository essays or written responses in the style of a specific writing genre, but there are many other types of writing opportunities for students to do throughout the Novel Engineering trajectory. Think about when you want students to write and your goals for each of the writing tasks. Table 12.1 includes different writing prompts and their place within the trajectory. We have combined the first and second steps—Read Book and Identify Problems—since students do these at the same time. This table is intended to show possible writing projects so you can pick which ones fit with your goals for your students.

In Table 12.2, we include examples of writing prompts for specific books we've seen classroom teachers include in their units. These prompts form part of an overall literacy component; they are not the only literacy piece students are expected to do.

Problem-tracking sheets can be as simple as a numbered list for students to list problems as they find them or a prompt for students to think more analytically about the problems they find. One of our favorite tracking sheets consists of columns to record what problems the characters are experiencing, who those problems affect and how, and what can be designed that might solve the

Table 12.1: The Novel Engineering trajectory and possible writing prompts

Step of the Novel Engineering Trajectory	Possible Writing Prompt
Read Book and Identify Problems	• Problem-tracking sheet • Empathy map
Scope Problem and Design a Solution	• Brainstorming sheet or sticky notes • Planning sheet • Product description (prior to building)
Test Solution and Get Feedback	• Testing results recorders • Process reflection prompt
Improve Solutions	• Redesign document
Share Culminating Activity	• Essays: narrative, descriptive, expository, persuasive • Presentations • Mix of visual and written communication

Table 12.2: Writing prompts for specific books

Book and Author	When Prompted	Writing Prompt
The City of Ember by Jeanne DuPra	After building is complete	Take the perspective of a character and write a diary entry on the new design and how it affects their life.
From the Mixed-Up Files of Mrs. Basil E. Frankweiler by E. L. Konigsburg	After building is complete	Write a new ending for the book if your design has changed the plot significantly. You may use the author's voice or tell it from the perspectives of Claudia, Jamie, or Mrs. Basil E. Frankweiler.
Hatchet by Gary Paulson	After building is complete	• Write a reflection on what was interesting and challenging in the design process. • Write a letter to the character (your client) about how he/she can use your design and how it would help him/her survive.
If You Lived in Colonial Times by Ann McGovern	After building is complete	• Write a story about living in the colonial era, incorporating your new design into your stories. You may describe how the new design changed ways of life for colonial people. • Write letters to children in the colonial era, describing how to construct the design with available materials.
The Invention of Hugo Cabret by Brian Selznick	During reading	• Keep a design journal in which you keep design notes, decisions, sketches, and other pertinent information. • Write a reflection about the design process thus far. Did you make any changes from your original plan? Why? How did these changes improve your design? Include a hypothesis and observations if you are testing or have tested your design. What changes did you make based on peer feedback?
James and the Giant Peach by Roald Dahl	After building is complete	Write a letter to James and the bugs about how the design works and how it will help them. Discuss materials they can use that are inside the peach or in the ocean.
A Long Walk to Water by Linda Sue Park	After building is complete	• Write to the author, Linda Sue Park, and Salva about your ideas. • Write a letter to Nya about how your design works and how it will help her and her family. Discuss materials they can use in Sudan or how they could get them.

(continued)

Table 12.2: Writing prompts for specific books *(continued)*

Book and Author	When Prompted	Writing Prompt
The Miraculous Journey of Edward Tulane by Kate DiCamillo	During reading	Oh no! Edward is overboard! What are you thinking? How do you feel? How would you save Edward? Pretend you're Abilene or Edward and write a diary entry about your feelings, thoughts, or plan of rescue.
The Mouse and the Motorcycle by Beverly Cleary	After building is complete	Write a reflection on how scale played into your design process.
Muncha! Muncha! Muncha! by Candace Fleming	After building is complete	Write a letter to Mr. McGreely about how your design will help him or why you chose to help the bunnies.
Pop's Bridge by Eve Bunting	While building or after building is complete	Write a letter to Robert telling him about your design and how it will protect his father.
The Snowy Day by Ezra Jack Keats	During planning	Draw and label diagrams of what you will build.
The Tarantula in My Purse by Jean Craighead George	During planning	In your planning document, describe the following: • How will your design have changed the family's experience with the ducklings? • How does your concept allow people to use their bathtub or shower? • How does your concept move the ducks as fast and easily as possible and give them the feeling they had in nature?
The Trumpet of the Swan by E. B. White	During reading	• Write about what it means to be without a voice, both figuratively and literally. • Keep a journal like Sam does about your process with the book and engineering.
Tuck Everlasting by Natalie Babbitt	During planning or after building is complete	• Create an advertisement for your prototype that would convince a character to buy it. • Make a comic strip showing what happened when your prototype was used in the story.
Weslandia by Paul Fleischman	During reading	• Write a "history" of Weslandia. • Research another civilization and write your research findings.

problems (see Table 12.3, Figure 12.2, and Appendix J, p. 242). With such a sheet, students can list the problems as they read and then use the other columns to facilitate discussions.

Empathy maps can spark discussions about character development and lead students to make inferences about the characters that will influence their own design decisions. Filling out an empathy map helps students better understand

Table 12.3: Problem-tracking and scoping sheet

What is the problem?	Who does this problem affect?	How does it affect them?	What could you design to solve the problem?

Figure 12.2: Sample problem-tracking sheet for *A Long Walk to Water*

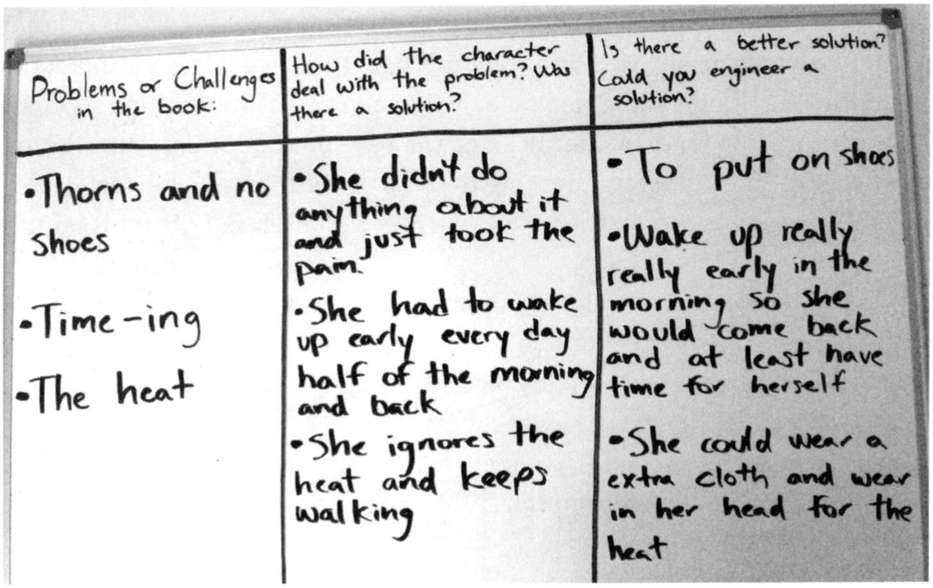

the characters, which can give them information about constraints and criteria of their design. These sheets can be filled out relatively quickly and provide students with data as they choose design criteria. For example, while reading *Tuck Everlasting*, students may sense that one of the characters, Winnie Foster, feels scared as she finds herself with a stranger. By filling out an empathy map as Winnie Foster, a pair of students can identify not only things Winnie does and says in the story but also what she might have felt and thought. Later, students can use what they identified on the empathy map to help them scope the problem they've chosen and find solutions. (See Appendix K, p. 244, for an example of a blank empathy map and an example of student work from *Tuck Everlasting*.)

Brainstorming can be documented on sticky notes, on a brainstorming worksheet or planning sheet, or in a journal. A sample brainstorming sheet is included in Appendix H (p. 235). The important thing to keep in mind about the brainstorming step is to keep it fairly open-ended so students are not constrained by a teacher's demands. Brainstorming should be about having students explore their ideas—not about meeting a requirement for ideas.

Planning documents can be simple or complicated, depending on the amount of time you have set aside for the unit, the experience level of your students, and your goals. Figure 12.3 shows a planning sheet used for the third graders who read *If You Lived in Colonial Times*. For older or more experienced students, you can use a more detailed planning document (see Appendix I, p. 236).

Worksheets can be used to help students keep track of the results of their design tests. These worksheets should be open enough to allow for different testing situations. Some groups may test their designs so frequently that a testing sheet would be cumbersome. The purpose of a testing sheet is to help students keep track of what worked well and what did not work well so they can analyze the results as they think about possible design changes. The testing recorders may not be the same for all students since they are building different solutions, but they should include what they were testing, the outcome of the test, and what they should change based on the results (See Appendix L, p. 246, for a sample worksheet).

Some teachers combine writing prompts into a journal so students can document the project as they go. These journals can be used for multiple projects so students collect all their work in one place and can reflect on their progress from project to project. Teachers who want their students to keep a journal should distribute a packet for each unit, such as the one included in Appendix I (p. 236), or have a blank notebook with similar information that can be used with projects throughout the year.

Figure 12.3: Student work for *If You Lived in Colonial Times*

Name_____Date_____.

Imagine you are an employee of "The Thirteen Colonies Tool Company." You have been hired by the colonists to create a tool to help them make life easier. Your group must come up with a plan to present to the colonists and once you have their approval, you must create the tool. You need to list all the materials that you will use, and come up with a price to charge for your time and product.

PROBLEM: If they were hot & cold

PLAN: bild wooden for the blows hot & cold air a

MATERIALS TO BE USED: sticks, coten ball tape a box

Assessing Student Work

Assessing students through their literacy tasks is a way to get at their under-standing of engineering, their design, and the text, and it can also provide clarity on how students worked through the engineering design process. Engineering journals can be used as part of your assessment. We know it would be impossible for students to complete a detailed writing prompt at every step of the process, so the written artifacts they create can be used as formative and summative assessments of their engineering work—and can give a more complete picture than just looking at the final product. The type and number of writing prompts you use should be chosen to address your writing goals and help students struc-ture their thinking.

Book Resources

Tuck Everlasting; Babbitt, N.; Age Range: 9–11; Lexile Level: 770L
If You Lived in Colonial Times; McGovern, A.; Age Range: 7–10; Lexile Level: 590L
A Long Walk to Water; Park, L. S.; Age Range: 10–12; Lexile Level: 720L

Implementing a Novel Engineering Unit in the Classroom

Chapter 13

This chapter covers how to implement Novel Engineering in the classroom. It will describe each step of the Novel Engineering trajectory and how to support students as they begin engineering in the classroom. A sample unit for *America's Champion Swimmer: Gertrude Ederle* by David A. Adler is available on the book's Extras page at *www.nsta.org/novelengineering*.

Introducing the Unit

If this is your students' first experience doing Novel Engineering, you will want to provide them with some context before you jump into an activity. As we've mentioned, starting with a whole-class discussion about what they will be doing helps orient students and gives them an understanding of what the expectations will be. (Chapter 9 discusses specific strategies and goals for introducing engineering and Novel Engineering to your students.)

Once students understand how they will be interacting with the text, you may decide to lead a discussion to establish class norms in engineering. Specifically, you'll want to help students understand how they will interact with one another and what it means to work collaboratively to build a working solution for a problem from the text. If you've done Novel Engineering before, it may be helpful to share a few examples of what past students have done, describing their process and how they worked together. Students benefit from hearing about things that did and did not go well, and showcasing functional solutions from a past classroom experience helps students understand what it means to have working, testable designs as part of their criteria.

Reading and Discussing the Book

There are many options for how students can read the book, such as individually, in literature circles (see Figure 13.1), or as a whole-class read-aloud. Schedule time for students to reflect on what they've read, identify key concepts, and think about and discuss the problems that have arisen in the book.

We encourage you to use literacy strategies that are familiar to you and your students, but use an "engineering lens" to dig deeper into the story. The more frequently you help students notice and discuss problems in the text, the more comfortable they will be at identifying and scoping problems themselves. These discussions also enable students to better understand the book's characters, setting, and themes, which, in turn, help them understand the constraints that will guide their designs.

Figure 13.1: Students reading *The Invention of Hugo Cabret* by Brian Selznick

Naming Problems

In general, we have found that it works well to stop reading at predetermined points, such as the end of a chapter, and ask students to reflect on and name the problems they noticed in that section of text. Teachers often add those problems to a large piece of chart paper or on the whiteboard to keep a running record of all problems students identify.

The age and experience of your students will influence how much support they need as they identify problems. Very young students in kindergarten and first grade may not be sure of what a problem is, so that may be the focus of your first engineering and literacy experience. If this is the case, you may want to read as a whole class and prompt students to identify problems as you read. Remind students that all problems—big and small—are important. If students seem unsure of what to do or get stuck trying to find the "right" problems, you can pick one or two problems from the text and discuss why they are problems. For older students, it is typically helpful for each of them to have a piece of paper or journal to jot down problems as they read. These lists can then be used when the class comes together for discussions. This is described in more detail in Chapter 12.

Teachers often ask, "What are appropriate problems for students to solve?" This is a difficult question to answer, especially when there is no way of knowing the types of designs students might create! Generally, problems that lead to in-depth discussions and an understanding of the text are open-ended and require students to make inferences about the characters. We've found that problems of this nature require students to revisit the text to uncover new details and discuss how those details might relate to the problem. For younger students, the process of identifying problems may need to be more structured.

There are also problems that feel more social in nature and cannot be solved by engineering. We encourage you to listen to your students' solution ideas before deciding that a problem is not appropriate for Novel Engineering. Having an explicit conversation with students about these two types of problems (social versus engineering) helps them make sense of what they are being asked to do and lessens their frustration.

After students have listed as many problems as possible, it is helpful to review the list and categorize problems as either problems that can be solved by engineering in the classroom or problems that cannot. The latter category might include problems such as "An airplane engine is broken." Even though this problem can be solved by engineering, it would be impossible to procure the necessary equipment in a classroom environment. Grouping problems into these two categories is beneficial for two reasons. First, it forces students to think about material and time constraints. Like professional engineers, they do not have unlimited time and resources. Second, it helps students understand what constitutes a feasible, functional solution. They learn to disregard problems that require advanced technical abilities or materials, as well as those that involve magic or fantasy. As you become more experienced with Novel Engineering, you will learn to anticipate what problems students will find, what types of solutions they may design for those problems, and the stumbling blocks they will likely encounter.

Brainstorming and Planning

We recommend putting students in small groups or pairs before they begin brainstorming and planning. In pairs, they can choose a problem from the class list and start brainstorming solutions. We also recommend that students take time to brainstorm solutions for three different problems. All ideas should be included in students' collective brainstorming; however, we don't discourage early considerations of feasibility for those ideas. For instance, you may hear students refer to the book or look at available materials as they pick a problem. Students should keep track of their brainstormed ideas by writing them down on paper, in their notebooks, on sticky notes, or on planning sheets.

After students have had time to brainstorm (typically 10–15 minutes, more if you hear productive discussions), they should have a chance to choose one solution to work on together. Of course, sometimes arriving at a consensus is not simple for students who feel strongly about their own ideas. Two students may have competing ideas or be unable to agree on a single idea. In these instances, we recognize that an adult may need to facilitate a discussion, so we recommend guiding students to practice listening strategies. When students give one another a chance to fully explain an idea without interrupting, the listening student may discover something he or she likes about the idea. Understanding one another's ideas also helps students brainstorm new ideas that draw on the strengths of both ideas. When students work together to come up with a new idea, they are more engaged because they feel some ownership over the shared idea.

When students struggle to agree on a solution, we do not typically recommend they keep both ideas and build in parallel. Part of what students are learning to do in this process is work collaboratively. We understand that collaboration can be extremely difficult for some students—perhaps it's even a goal on an Individualized Education Program. If this is the case, identify roles or parts of a design that will allow students to work independently on some tasks while collaborating on others.

As students move from brainstorming to planning, their ideas become more solidified and they begin to think more seriously about materials. Even though they will not test until later, we recommend letting students know that testing is an expectation so they have it in mind as they plan. For some students, this keeps them grounded in building functional solutions. Other students may need to be reminded several times that they are expected to build something that works. Many teachers simply ask, "How will it work?" or "How are you going to show us that it works?" Students should be aware of which materials will be available to them while they are planning. It's difficult to plan a solution if you don't know what materials you have and how they will work together.

We recommend students keep all their planning ideas in an engineering journal or on a worksheet. This will help students structure their thinking and document their process. It may also help them communicate their ideas to other people. A basic planning document should include the problem that is being solved, the proposed solution, how it will work, and required materials. (Examples of student work are included in Appendix I, p. 236). More detailed planning documents can include design constraints and criteria. The amount of detail included usually depends on the age of the students, the number of times they have previously planned a Novel Engineering project, their abilities, and your writing goals for them. Younger students and those who have not participated in an engineering activity will need guidance regarding the level of detail you are looking for in the planning document. It's fine if not everyone's planning document has the same level of detail; Novel Engineering allows teachers to differentiate for specific students' needs while allowing equitable participation in the unit.

While planning, students will continue to develop their ideas as they communicate with others, work with different materials, and research similar solutions that may already exist. If there is time, we recommend having a quick share-out where each group can share their idea (possibly in a two-minute pitch) and classmates can ask questions (typically one to three questions). The planning stage is an opportune time for sharing early ideas because students have not yet started building and may be more flexible with the development of an idea. Students may also provide and receive suggestions that help their classmates figure out how they will make a design work or how to test it.

Encouraging Solution Diversity

One of the goals of professional engineers is to design novel or new solutions. We encourage the same for young students. You can support this by allowing students to follow their individual ideas and encouraging them to listen to their classmates' ideas. Providing students with a variety of materials allows them to create diverse solutions. If material choice is sparse or students are all given the same kit with a limited selection of materials, it is harder for them to be creative. When a wide variety of materials is available, students rarely want the same things—with the exception of tape!

Building and Prototyping

Once students have outlined their plan, they can gather materials and begin building. A good way to make sure students are ready and have not planned an impossible solution (e.g., a shrink ray device) is to tell them they need to check with you first. This will ensure that you understand what each group is proposing and the type and quantity of materials they want. It will also allow you to make sure that one group is not monopolizing a certain material. If you feel this is the case, talk to students and discuss if another material could also be appropriate.

The aspect of building that teachers ask about most is how to hand out materials equitably. (See Chapter 10 for more information about choices related to materials.) One way is to give all groups the same materials. This is certainly equitable, but it may hamper students' creativity and lead to a lack of diversity among the solutions. Students may also spend time negotiating trades that will take time away from building. Most teachers show the available materials as students are planning and allow them to get their own materials once their plans have been approved. In general, students stick to their approved materials list. Sometimes, as students' designs evolve, they need different or additional materials. This is completely expected and shows that they are iterating on their designs.

We suggest telling students they need to have their initial plans approved in order to get their materials. This will ensure that students are getting materials that are appropriate to implement their plan. This also allows teachers to question students about the amount of materials they are asking for. As students redesign and ask for additional materials, they can be told that they must justify the need for those materials. Including these check points will help ensure that students are not randomly getting materials or taking more than they need.

Giving and Getting Feedback

Feedback gives students information that can be used to improve their designs. There are several opportunities during the engineering design process (EDP) for students to get feedback on their designs (see Figure 13.2). The first is to have a brief "design pitch" while students are still planning. With a design pitch, the

Figure 13.2: Types of design feedback

Name __Mike__   Partner(s) __Thomas__

The problem we are going to design a solution for is __how to see Angel__

We chose this as our problem because __Jammie and Claudia need to get a better look at Angel to see who sculpet her.__

Our plan: __Our plan is to build a periscope-like device to see above the heads of adults and around obstacles. But the mirrors could rotate.__

Sketch of our Design

Cardboard nods

cardboard body   Picture hooks

Materials we anticipate needing (Include how much of each material you think you will need):

- 5 medium sized cardboard packing boxes
- 2 6 by 6 cm mirrors
- 6 picture hooks
- 5 cm of stiff plastic tubing
- 2 roll duct tape – reel
- 4 thick rubber bands

teacher and classmates give initial feedback to make sure students are going down a path toward a functional, attainable design. A second opportunity is to allow for peer design reviews during a mid-design share-out. Students will be able to make changes to their designs based on the feedback they get. This is preferential to getting suggestions during a final presentation since they still have time to make changes. A third opportunity is to allow feedback while or after students have tested. Outside these times, students can informally give and get feedback by talking to teachers and peers as they build. The goal of all feedback is for students to think differently about their designs and make appropriate changes.

Testing

Most students move fluidly from testing to revising as they observe and gather test results. Testing should help students gather information about how successfully their designs function and how well they meet the established criteria. Ideally, students are conducting small tests as they work on different components of their designs. The data students collect will be qualitative or quantitative, depending on the design and testing situation. Testing results should be evaluated based on design constraints and criteria that were identified during the planning stage.

The teacher's role during this step is to help students recognize that testing is a way to gather information about what is working well and what needs to be altered. It should not be seen as a negative reflection of their work. In fact, as students test more frequently, they begin to see testing as a valuable source of feedback. They also become more skilled at analyzing and interpreting data.

Tests should be authentic so they help students better understand their designs and measure specific design criteria. In a unit based on *Peter's Chair* by Ezra Jack Keats, kindergarten students built chairs for the main character and used a doll to test them. The doll gave them information about the size and sturdiness of their chairs. By contrast, in a unit on *The Miraculous Journey of Edward Tulane* by Kate DiCamillo, a plastic tub was filled with water to represent the ocean. This test proved inauthentic because the sides of the tub did not realistically simulate the ocean floor. As a result, this test did not provide appropriate information to let students know if their designs would work on the ocean floor.

In addition, tests should be responsive to students' designs. If each group is addressing the same problem, it makes sense for the teacher (or the class) to come up with a way to test their solutions. However, if each group is addressing a different problem or building vastly different solutions, each test will be uniquely developed for the specific design solutions. For example, in one third-grade classroom reading *If You Lived in Colonial Times*, the tests for students building a water filter, a medicine grinder, and a digging tool were completely different. For each of these designs, students had to come up with a test to measure whether the solution met the design criteria. To help students get into the appropriate mindset to test the functionality of their designs, we often ask, "How will you know that it works?" or "How will you show us that it works for (character)?"

Conducting Peer Reviews

Another form of feedback is a structured conversation for students to share their ideas with their peers. These design reviews, or share-outs, can happen at several times during the Novel Engineering design process. Although final presentations are a traditional part of design work, we've found it more beneficial for students learning about the design process to have a share-out before they complete their designs. Mid-design share-outs reflect the process professional engineers follow when working in a team or pitching design ideas, and they give students the opportunity to receive comments and advice before they complete their designs.

As students participate in peer reviews, shared agreements of a successful design emerge, and they begin to hold one another accountable to design constraints and characters' needs. In addition, students become more skilled at giving and taking advice and more comfortable with the design process. The feedback students receive from their peers in a mid-design share-out often affects their designs, allowing them to flexibly make changes as they work. When an evaluation session happens after building is complete, the conversations are much different since students will not be making changes.

Before students address the class, it is important for them to know what is expected in a design share-out. Let them know they will be sharing their ideas with the whole class and they should be prepared to demonstrate how their designs work, how the designs solve the chosen problem, and what is not going well (so they can get ideas from their classmates). Stress to students that their job when not presenting is to listen to the presenting group so they are able to ask pertinent questions and offer suggestions. Ideally, students will have a few sessions to build before the share-out. A mid-design share-out may take 20–40 minutes for the entire class, depending on the number of students and how long they talk. After sharing, students should have at least 20–40 minutes to make improvements and finish their designs.

You may also need to explicitly address how to give useful feedback. Critiques should go beyond saying that you like a design or that it looks nice. Suggestions should help the presenting group assess what is working and what needs to be improved. Ron Berger, the founder of EL Education, outlines three rules for classroom critique that are simple enough for students to remember (Berger 2016).

1. Be kind.
2. Be specific.
3. Be helpful.

It is important to provide students with sufficient time to revise or improve their designs based on feedback from tests and classmates. Redesign is crucial to understanding how to problem solve, especially in engineering. Allowing students to grapple with problems and elicit help from peers also fosters more positive attitudes around failure; students recognize a failure in their design as an opportunity to get feedback and improve their work.

Students (and teachers) should understand that a design is never "done." It can always be further optimized. If engineers had unlimited time and resources, they would be able to continually improve designs. However, in most cases, engineers are also dealing with project schedules and tight budgets. A design is "done" when an engineer has made it the best it can be within the given constraints. The same is true in the classroom. At some point, sometimes even before a design is functional, students need to stop working. It is therefore crucial to celebrate the work students have completed—the ideas and prototypes they've had—and their design process.

By contrast, you may find that other students think they are done even when the design is not functional or can be modified to better meet the criteria. When students tell you they have completed their design, ask them to show you how it works. You may find, for example, that a student needs to hold part of the structure to make it work or that it is not stable enough to stay together. If students are unable to make their design work or if success is not repeatable, encourage them to spend more time on it so it is consistently successful.

Working With Time Constraints

As much as unlimited time would be wonderful in terms of both student engagement and learning, this is simply not the reality of a classroom. When you begin a unit, it is important to let students know expectations and how much time they will have to complete it. It also helps to remind students that their design process and understanding of the story are just as important as the final product. This will help alleviate students' stress if they do not finish building their design in time. Being clear with students about the allotted time and check-in points along the way helps them organize their work and manage their time. If, during the project, you become concerned that some students may not complete their project as planned, help them narrow their building focus so they are successful with a portion of the project.

Designs often have multiple components, so there may be several working parts to students' designs they need to address as they build. Students need to build and test each of these components. However, if it becomes clear that students may not finish building all the parts of a design, it is important to help them prioritize the parts that need to work in order to solve the problem. It's important for students to think about the order in which they will complete things. The ability to manage time and organize workflow is a learned skill. If students are young or have not had many experiences managing their time for larger projects, they may benefit from a discussion about priorities and timelines, preferably during the planning stage of the project. For example, you could discuss with students that, in the real world, engineers generally do not work on all parts of a product. Rather, there are teams of engineers; each focuses on a different component, and they work together to make the whole design.

Sharing Final Designs and Wrapping Up

There are several ways for students to share their final designs and reflect on the unit. These include final presentations, writing tasks, and less structured sharing sessions. Often, teachers plan for a final presentation and culminating writing activity so students can showcase what they built, how it relates to the story, and how it addresses the characters' needs.

Since not all students will finish at the same time, there are a few tasks they can do if they finish early. For example, they can act as a photographer or a videographer and help other students document their work. If you are including a final writing assignment as part of the unit, students can begin to work on that assignment. Student can also reflect on and outline their unique design processes with a mobile app or on paper.

If you choose for students to do final presentations, the presentations should focus on the process, how the design meets the needs of the client(s), and how the final product works. In describing their process, students have an opportunity to reflect on and share the evolution of their designs—how the designs changed over time, what happened when they tested specific components, and how they improved their designs. This helps students understand how iteration is important and how failure is a way to assess what is not working in a design. Final presentations are also an opportunity for whole-class reflection and for the teacher to assess students' learning.

It's helpful to discuss presentation norms before you begin, especially if students have not had experience giving or watching presentations. For instance, you might have a brief discussion about what makes an effective presentation, focusing on qualities such as talking loudly and clearly addressing the main points. You might also discuss the qualities of an effective listener, including making eye contact and not talking while someone else is presenting. These ideas help students develop expectations and norms for presenting in a classroom environment. The goal is for each student to take ownership of his or her presentation and lead a discussion about the solution.

Students should understand that, although a "final" design in being presented, this is a chance for them to acknowledge aspects of their design that could be improved if they had more time or resources. Students might mention what they would change if given the opportunity, or describe the limitations of the materials they used. This may prompt students to describe what their design would look like in the "real world" if they had access to more time and different resources. If students built their design out of cardboard, for instance, they might explain that it would ideally be made out of another material, such as wood or metal. This helps students see that design is an ongoing process and that a successful product may require several iterations and representations. If there is not enough time for a mid-design and a final share-out, we recommend the mid-design share-out since feedback is such an integral part of the EDP.

If you are concerned that there might not be sufficient time for final presentations, there are other ways that students can share and see one another's work. A gallery walk is an activity where students walk around and look at and discuss one another's designs. This is quicker than sharing and discussing each group's design, but it still gives students the chance to see one another's completed projects. Having students take short videos of their designs and then presenting them as a video showcase achieves the same goal.

Looking Beyond Novel Engineering

Since Novel Engineering is so flexible, it can be structured to meet many different types of goals—academic, social, and interpersonal—and give students experiences in engineering and literacy that feel personally meaningful and encourage them to think about the needs and desires of others. Participation in Novel Engineering is not the final destination. These learned skills can become the foundation of future efforts, both in and out of school. The scope of projects can move beyond books to looking at real-world problems in the classroom, school, or community. It is our hope that students will build on the skills they acquire during Novel Engineering to analyze and address problems, both big and small, they see in the world.

Working on this project has been a fulfilling experience due to the excitement and engagement of the teachers and students involved. We hope that as you implement Novel Engineering with your students, you get to observe the same energy, conversations, and creative student work that we've witnessed.

Safety Notes

1. Wear safety goggles or glasses with side shields during the setup, hands-on, and takedown segments of the activity.
2. Use caution when using hand tools that can cut or puncture skin.

Reference

Berger, R. 2016. "Austin's Butterfly: Building Excellence in Student Work." video file. *https://eleducation.org/resources/austins-butterfly*.

Website

EL Education: *www.eleducation.org*

Book Resources

James and the Giant Peach; Dahl, R.; Age Range: 7–10; Lexile Level: 870L

The Miraculous Journey of Edward Tulane; DiCamillo, K.; Age Range: 8–11; Lexile Level: 700L

Peter's Chair; Keats, E. J.; Age Range: 3–7; Lexile Level: 500L

From the Mixed-Up Files of Mrs. Basil E. Frankweiler; Konigsburg, E. L.; Age Range: 8–12; Lexile Level: 700L

If You Lived in Colonial Times; McGovern, A.; Age Range: 7–10; Lexile Level: 590L

A Long Walk to Water; Park, L. S.; Age Range: 10–12; Lexile Level: 720L

The Invention of Hugo Cabret; Selznick, B.; Age Range: 9–12; Lexile Level: 820L

Appendixes

Novel Engineering Materials List

Suggested List

Recyclables	Things in Classroom/Home	To Purchase
Paper tubes	Construction paper	Wooden dowels
Cardboard	Trash bags	Bamboo gardening sticks
Egg carton	Dead batteries	Pipe cleaners
Coffee stirrers	Craft sticks	Wire
Paper	Index cards	Plastic tubing
Newspaper	Glue	Chipboard
Poster board	Paper clips	Wax paper
	Rubber bands (nonlatex)	Cork sheet
	Straws	Transparency film
	Plastic silverware	Cloth
	Toothpicks	Quilt batting
	Coffee filters	Netting or mesh
	Paper plates	Fishing line
	Tin foil	Magnets
	Duct tape	Zip ties
	Masking tape	Fastening tape
	String	Modeling clay
	Thread	Pulleys
	Cotton balls	Painting tape
	Paper cups	Electrical tape
	Packing material	Clothespins
	Felt	Plastic food containers
	Foam	Sand
		Gravel

Connections to *Next Generation Science Standards*

Next Generation Science Standards		Novel Engineering Arc					
		Read Book and Identify Problems	Scope Problems and Brainstorm Solutions	Design Solutions	Get Feedback	Improve Solutions	Reflect and Share
K–2 Standards	K-2-ETS1-1. Ask questions, make observations, and gather information about a situation people want to change to define a simple problem that can be solved through the development of a new or improved object or tool.	X	X				
	K-2-ETS1-2. Develop a simple sketch, drawing, or physical model to illustrate how the shape of an object helps it function as needed to solve a given problem.		X	X		X	X
	K-2-ETS1-3. Analyze data from tests of two objects designed to solve the same problem to compare the strengths and weaknesses of how each performs.				X	X	
3–5 Standards	3-5-ETS1-1. Define a simple design problem reflecting a need or a want that includes specified criteria for success and constraints on materials, time, or cost.	X	X				
	3-5-ETS1-2. Generate and compare multiple possible solutions to a problem based on how well each is likely to meet the criteria and constraints of the problem.		X	X		X	X

(continued)

Connections to *Next Generation Science Standards (continued)*

Next Generation Science Standards		Novel Engineering Arc					
		Read Book and Identify Problems	**Scope Problems and Brainstorm Solutions**	**Design Solutions**	**Get Feedback**	**Improve Solutions**	**Reflect and Share**
3–5 Standards *(continued)*	3-5-ETS1-3. Plan and carry out fair tests in which variables are controlled and failure points are considered to identify aspects of a model or prototype that can be improved.				X	X	
6–8 Standards	MS-ETS1-1. Define the criteria and constraints of a design problem with sufficient precision to ensure a successful solution, taking into account relevant scientific principles and potential impacts on people and the natural environment that may limit possible solutions.	X	X				
	MS-ETS1-2. Evaluate competing design solutions using a systematic process to determine how well they meet the criteria and constraints of the problem.		X	X		X	
	MS-ETS1-3. Analyze data from tests to determine similarities and differences among several design solutions to identify the best characteristics of each that can be combined into a new solution to better meet the criteria for success.				X	X	X

Connections to *Common Core State Standards*

Common Core Standards **English Language Arts Literacy**		**Novel Engineering Arc**					
		Read Book and Identify Problems	**Scope Problems and Brainstorm Solutions**	**Design Solutions**	**Get Feedback**	**Improve Solutions**	**Reflect and Share**
Reading	CCRA.R.1. Read closely to determine what the text says explicitly and to make logical inferences from it; cite specific textual evidence when writing or speaking to support conclusions drawn from the text.	X			X		X
	CCRA.R.2. Determine central ideas or themes of a text and analyze their development; summarize the key supporting details and ideas.	X					X
	CCRA.R.3. Analyze how and why individuals, events, or ideas develop and interact over the course of a text.	X	X				X
Writing	CCRA.W.1. Write arguments to support claims in an analysis of substantive topics or texts using valid reasoning and relevant and sufficient evidence.						X
	CCRA.W.2. Write informative/ explanatory texts to examine and convey complex ideas and information clearly and accurately through the effective selection, organization, and analysis of content.						X
	CCRA.W.3. Write narratives to develop real or imagined experiences or events using effective technique, well-chosen details and well-structured event sequences.						X

(continued)

Connections to *Common Core State Standards (continued)*

Common Core Standards **English Language Arts Literacy**		Novel Engineering Arc					
		Read Book and Identify Problems	Scope Problems and Brainstorm Solutions	Design Solutions	Get Feedback	Improve Solutions	Reflect and Share
Writing (continued)	CCRA.W.6. Use technology, including the Internet, to produce and publish writing and to interact and collaborate with others.						X
	CCRA.W.9. Draw evidence from literary or informational texts to support analysis, reflection, and research.			X	X		X
	CCRA.W.10. Write routinely over extended time frames (time for research, reflection, and revision) and shorter time frames (a single sitting or a day or two) for a range of tasks, purposes, and audiences.			X	X		
Speaking and Listening	CCRA.SL.1. Prepare for and participate effectively in a range of conversations and collaborations with diverse partners, building on others' ideas and expressing their own clearly and persuasively.	X	X	X	X		
	CCRA.SL.2. Integrate and evaluate information presented in diverse media and formats, including visually, quantitatively, and orally.			X			
	CCRA.SL.3. Evaluate a speaker's point of view, reasoning, and use of evidence and rhetoric.			X	X	X	
	CCRA.SL.4. Present information, findings, and supporting evidence such that listeners can follow the line of reasoning and the organization, development, and style are appropriate to task, purpose, and audience.		X	X	X	X	

(continued)

Connections to *Common Core State Standards (continued)*

Common Core Standards **English Language Arts Literacy**		Novel Engineering Arc					
		Read Book and Identify Problems	**Scope Problems and Brainstorm Solutions**	**Design Solutions**	**Get Feedback**	**Improve Solutions**	**Reflect and Share**
Speaking and Listening (continued)	CCRA.SL.5. Make strategic use of digital media and visual displays of data to express information and enhance understanding of presentations.						X
	CCRA.SL.6. Adapt speech to a variety of contexts and communicative tasks, demonstrating command of formal English when indicated or appropriate.		X	X	X		

Product Comparison Worksheet

Group Members: _____

Item	Criterion 1	Criterion 2	Criterion 3	Criterion 4

Properties of Materials Worksheet

Material〳Properties				
Strength: • How hard can you pull on it? • How much weight does it hold? • Can you push it together? What happens? • How can you break/tear it?				
Ability to React to Air: • What can you do to make it blow away farther? • What can you do to it to keep it from flying away?				
Ability to React to Water: • When does it float? • How can you change it to make it sink? • Is it waterproof? Can you make it waterproof?				
Ability to React to Joining or Separation: • How can you connect it? • Does it pull apart easily?				

Novel Engineering Unit Planning Document

1. NAME OF BOOK: _____	**THINK ABOUT …**
Notes and Stopping Points:	• Why do you think this book will work? Why did you choose this book? • Does this book have multiple problems that can be solved by engineering? • Does this book have details about the character and setting? • Will you read the whole book or will you stop at a specific point (e.g., where a character has a problem but before it's solved)?
2. PROBLEMS	**THINK ABOUT …**
What problems will students identify? Pick one of the problems you've identified and think of multiple solutions to the problem.	• Are there multiple problems that your student can identify? • Are there multiple ways to solve the problems with engineering? • *Reconsider your book choice if you have a hard time thinking of multiple problems and solutions.*
3. MATERIALS	**THINK ABOUT …**
What materials will you need? How and when will you introduce students to materials?	• Where will you store materials? • How will you allocate materials? • Depending on their experience, students may need time to play or experiment with materials. • Will you give all students/groups the same materials or will there be choice?
4. ENGINEERING	**THINK ABOUT …**
What constraints will you impose, if any? How will you model communication of design ideas and constraints? What protocols might you consider using to share ideas and designs in process?	• How will you help students consider multiple solutions? • How will you help them work toward functional designs? • When would it be appropriate to have a model versus a functional design? • How will you provide space and time for the testing of students' designs? • Which engineering habits of mind might you choose to focus on?

(continued)

Novel Engineering Unit Planning Document *(continued)*

5. READING	THINK ABOUT ...
How will students read the book? How will you or students capture the problems identified? How will students understand the characters well enough to design for them?	• Reading options (whole-class interactive read-aloud, individually, reading groups) • Capturing problems (anchor chart, note catcher, stop and jot, reading journal) • Understanding the character
6. WRITING	**THINK ABOUT ...**
What writing assignments will be incorporated into the unit/activity?	Possibilities include the following: • Reflections on process • Comic strip • Alternative ending • Additional book chapter that includes the design • Planning documents • Instructions to character to use or build design • Presentations
7. MANAGEMENT	**THINK ABOUT ...**
How will students be grouped for building? What worksheets do you need?	• Will the entire class work on the same problem or multiple problems? • How will students plan?
8. ASSESSMENT	**THINK ABOUT ...**
How will you assess students?	• Consider whether you will assess individuals or groups. • What evidence/work will you look at? • Which standards will you assess (reading, writing, group work, science, engineering)?

(continued)

Novel Engineering Unit Planning Document *(continued)*

9. TIMELINE (Session 1, Session 2, Session 3 …)	THINK ABOUT …
	• How will you introduce Novel Engineering? • Elements to plan for 　— Reading the book 　— Discussing and identifying the problem 　— Planning 　— Introducing materials 　— Building and testing 　— Mid-design sharing 　— Redesigning 　— Final Sharing 　— Writing 　— Evaluating/Assessing

Sample Letter for Requesting Materials

Date

Dear Parents,

This school year, your son/daughter will take part in a pilot program called Novel Engineering. This program, developed by Tufts University's School Center for Engineering & Outreach, is an integrated approach to teaching STEAM-based literacy where students can design and develop projects based on texts they read in class. The characters in the story become the students' clients, and the student-engineers pull from the text to identify problems, set constraints, and design solutions.

Students will be guided to work on complex problems using the engineering design process. This process will allow students to listen, understand, and articulate their thinking while they explore different solutions to the problems. We will work with students to support their engagement and coach them to build on their ideas.

We will be collecting various building materials for our engineering design space for this pilot. Listed below are some of the items we will need. If you would like to contribute to our collection, please send any of the listed items to school with your child:

- Duct tape (gray)
- Plastic straws
- Paper clips
- Fishing line
- Wooden skewers
- Toothpicks
- Coffee stirrers
- Craft sticks
- Paper cups

- Pipe cleaners
- Construction paper
- Empty cereal boxes
- Plastic food containers (if used, please make sure they are rinsed well)
- Binder clips
- Empty tissue/shoe boxes
- Empty toilet paper/paper towel rolls

We are very excited to pilot this approach that integrates both English language arts and STEAM concepts. To find out more information about Novel Engineering, please visit *www.novelengineering.org*.

Sincerely,

Brainstorming Worksheet

For this step, you must consider all the possible solutions for your problem. Everyone's ideas are welcome, even if they seem wild. Please record them below.

What problem are you solving?

What are possible solutions to this problem? (List all the possible solutions you can think of.)

Planning Worksheet

Name: _____ Partner(s): _____

The problem we designed a solution for was _____

We chose this as our problem because _____

Problems we encountered or changes we had to make (how did your actual prototype differ from your original design?) _____

Explain to the character(s)—through step-by-step instructions—how to make your prototype.

Include a final list of materials: _____

How would having this invention change the book? _____

Draw a picture of the character(s) using your invention in a specific scene from the book.

```



```

Sample Planning Worksheets

Include a final list of materials.

① 2 medium-sized cardboard boxes

② 1 role of masking tape

③ 2 14 cm × 10 cm mirrors

④ 1 foam mirror frame

How would having this invention change the book?

This invention would change the book by setting a shorter timeline in the book, and give them more time to find the M and get the letter to the museum before he found out about it.

Draw a picture of the character(s) using your invention in a specific scene from the book.

Dream: This step you must consider all of the possible solutions for your problem. In this step, **everyone's ideas are welcome** even if they seem wild. Please record them all below.

- wristband- flips open - erasable - fabric elastic bands
- seceret code
- latch (could go with wristband)

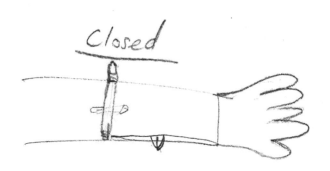

Name:_____

Partner(s):_____

The Mouse & The Motorcycle Design Challenge

1) What problem are you working on?

Stuck in the trash can

_____

2) How will it help solve the problem?

Make a . rope ladder and rock climb up

dresser.

_____

3) Draw a picture of what you will build. Label any important pieces and give the design a title. (You can use the back of this page.)

Your name: _____   Your partner: _____

Earthquakes!

Earthquakes can cause many different problems. Some of the problems are very serious. Other problems are minor.

Describe the problem you decided to solve:

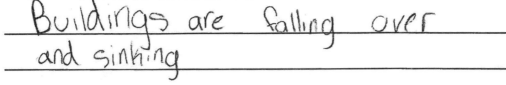

Buildings are falling over and sinking

Make a drawing of your solution.

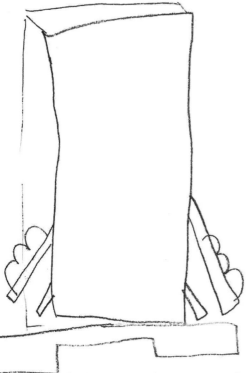

Explain your solutions in words:

We are making girders that are heavy and hard so it will not sink.

Problem–Tracking Worksheet

Name:

Problems or challenges in the book:	How did the character deal with the problem?	Is there a better solution? Could you engineer a solution?

Sample Problem-Tracking Worksheet

Name:

Problems or Challenges in the book:	How did the character deal with the problem? Was there a solution?	Is there a better solution? Could you engineer a solution?
Thorns	She tried taking the thorn out with another thorn.	She could try pushing the thorn out.
water	She just drank the dirty	Boil the water at a count to 200.
sister sick	She took her to the Doctor.	there is no better solution
sister	Nia had to walk slow for her sister.	?
No shoes	there was no solution.	I could make a type of shoe/flipflop for Nia.
Heavy water	She put the heavy water on her head	A type of tranportation Device

Empathy Map Worksheet

Character: _____

Says	Does

Thinks	Feels

Sample Empathy Map Worksheet

Character: <u>Whinnie foster</u>

Empathy Map

Marchesa

Says	Does
"No!" I won't go with you!"	• hugs tuck and keeps her eyes shut.
"I don't know his name"	
"They didn't kidnap me"	

Thinks	Feels
Wants to stay with the tuck family.	• Scared
Whinnie thinks that the Straser is bad.	• she feels that the tuck family is nice. • comphused about the spring. worried about the constable

Testing Sheet

Test #1	
What happened?	
Does this meet your criteria?	
What change(s) can you make?	

What are you testing?

Test #2	
What happened?	
Does this meet your criteria?	
What change(s) can you make?	

Index

Page numbers printed in **boldface type** indicate information contained in tables or figures.

Index

Index

Index

W

water filter, colonial times. *See If You Lived in Colonial Times* (McGovern) case study

Weslandia (Fleischman), **172, 202**

wheelchairs, 18–21

Winter, Amos, 18–21

Wolfson, Bill, 11

Wonder (Palacio), 5–7

workflow, 217

writing

about, 35

and assessment, 88–89

and documentation, 199–204, **200, 201–202, 203, 205**

and the EDP, 37–38

prompts for specific books, **201–202**